Cyber

Targeting America, Our Infrastructure and Our Future

Larry Bell

Cyberwarfare: Targeting America, Our Infrastructure and Our Future

Other books by Larry Bell:

Scared Witless: Prophets and Profits of Climate Doom
Climate of Corruption: Politics and Power behind the Global Warming Hoax
Cosmic Musings: Contemplating Life beyond Self
Reflections on Oceans and Puddles: One Hundred Reasons to be Enthusiastic, Grateful and Hopeful
Thinking Whole: Rejecting Half-Witted Left & Right Brain Limitations
Reinventing Ourselves: How Technology is Rapidly and Radically Transforming Humanity

STAIRWAY PRESS—Apache Junction

Book Cover Art: Masha Esfandabadi
Cover Design by Chris Benson
www.BensonCreative.com

STAIRWAY≡PRESS
www.StairwayPress.com
1000 West Apache Trail, Suite 126
Apache Junction, AZ 85120

Dedicated to those who work to defend us against threats we can hope never to learn about too late.

About the Author

LARRY BELL IS an endowed professor of space architecture at the University of Houston where he founded the Sasakawa International Center for Space Architecture (SICSA) and the graduate program in space architecture. Professor Bell has authored more than 600 column opinion articles for Forbes and Newsmax magazines, along with several books.

This book extends and expands the theme and contents of two previous books: *The Weaponization of AI and the Internet: How Global Networks of Infotech Overlords Are Expanding Their Control Over Our Lives* (2019), and *Reinventing Ourselves: How Technology is Rapidly and Radically Transforming Humanity* (2019).

Larry's other recent books include: *Thinking Whole: Rejecting Half-Witted Left & Right Brain Limitations* (2018), *Reflections on Oceans and Puddles: One Hundred Reasons to be Enthusiastic, Grateful and Hopeful* (2017*)*, *Cosmic Musings: Contemplating Life Beyond Self* (2016), *Scared Witless: Prophets and Profits of Climate Doom* (2015), and *Climate of Corruption: Politics and Power Behind the Global Warming Hoax* (2011). He is currently working on a new book co-authored with Buzz Aldrin, *Beyond Footprints and Flagpoles.*

Larry is also an entrepreneur who has co-founded several U.S. and international commercial space companies. One—established with NASA's Chief Engineer and two other partners—grew through mergers and acquisitions to employ more than 8,000 professionals, went on the New York Stock Exchange, and was purchased by General Dynamics.

Professor Bell's many national and international honors include two of the highest awards granted by Russia's prestigious Institute of Aeronautics and Astronautics: The Konstantin Tsiolkovsky Medal (Russia's "father of space rocketry"); and the Yuri Gagarin Diploma (for contributions to international space development.) Larry's name was placed in large letters on the Russian rocket that launched the first crew to the International Space Station.

Preface

THIS BOOK IS the third in a previously unintended series investigating ways that an information revolution is forever impacting human society both for the better and perplexing worse.

The first book, *Reinventing Ourselves: How Technology is Rapidly and Radically Transforming Humanity* (2018), addresses numerous and diverse ways that advancing information and computational technologies are dramatically impacting our lives, careers and societal values.

My second book, *The Weaponization of AI and Internet: How Global Networks of Infotech Overlords Are Expanding Their Control Over Our Lives* (2019), discusses consequential dangers as we constantly trade away more and more personal privacy and autonomy in exchange for promises of increased technological convenience and security.

This book, *Cyberwarfare: Targeting America, Our Infrastructure and Our Future* (2020), identifies and elaborates previous examples and current actions involving American cybersecurity forces, allies and adversaries that pose imminent national and personal security dangers.

What, exactly, is cyberwarfare?

The answer depends a great deal upon whom you ask, whom you blame, the extent of the damage and what retribution you are prepared to risk in responding.

For starters, let's presume that by "cyber war," we are generally referring to foreign (or even domestic) government-level nation-state actions that are intended to penetrate another nation's computers or networks for purposes of causing physical equipment damage or critical service disruptions.

In 2009, the U.S. government convened the National Research Council which defined cyberattacks as "deliberate actions to alter, disrupt, deceive, degrade, or destroy computer systems or the information and/or programs resident in or transmitting these systems or networks."[i]

Marie O'Neill Sciarrone, the Founder & President at Trinity Cyber LLC, emphasizes that cyberwarfare should not be thought of as a computer against computer, but rather as a much broader concept. It represents an effort through cyberspace or using digital means to attack an opponent for a variety of different purposes. Such objectives can range from state-sponsored infiltration with intent to disrupt information services to individual hackers trying to make a political statement or influence outcomes.[ii]

Such attacks can move at the speed of light, unlimited by geography and the political boundaries. Being delinked from physics also means it can be in multiple places at the same time, meaning that the same attack can hit multiple targets at once.

Cyberattacks often differ from traditional attacks because instead of causing direct physical damage, they virtually always first invade another computer and steal the information within it. The intended results may be to damage something physical, but that damage always first results from an incident in the digital realm.

Government-directed and employed cyberwarriors often don't act alone under their own national flags, but enlist and draw upon lots of assistance from covert surrogates including "patriot" hackers who afford cover for government deniability, social media cyberterrorists and independent cybercriminal elements. All of these groups apply common stealth strategies which make it difficult or impossible—and often irrelevant—to attribute actual motives or to fix blame.

Accordingly, the scope of this book will extend the definition of cyberwarfare to include references and discussions of dangerous actors who are often not officially linked to sanctioned government activities, yet who use the same tactics and tools. Whether those agendas are primarily political or economic in nature, the consequences can be similarly and expansively devastating.

Military and civilian intelligence organizations, for example, routinely prepare cyber battlefields with virtual computer explosives in the form of software malware called "logic bombs" and "trapdoors" placing virtual explosives in other countries' power grids, financial and communications networks, and other critical utility infrastructures.

Yes, and without any doubt, just as other governments do this to us, our cyber intelligence warriors do exactly the same to them. A big problem in our case, however, is that those nations with the greatest electronic infrastructure dependence also present richest, most vulnerable, cyberwar targets.

Every substantial cyberattack by one nation upon another provides a fair game precedent for potentially escalating see-saw levels of back-and-forth retaliation. A very considerable American disadvantage in this regard comes with living in the glassiest house in a very low-entry-cost stone fight.

Former White House security advisor Richard Clarke warns that at the very same time the U.S. prepares for offensive cyberwar, our current, and continuing policies simultaneously make it increasingly difficult to defend against cyberattack by aggressors. This is because we have simply too many vital networks to protect, and because additional ones are growing too quickly.

In simple terms, global offensive developments are wildly outpacing defense.[iii,iv]

David Sanger, who like Richard Clarke, teaches national security policy at Harvard's Kennedy School of Government, emphasizes fundamental differences between nuclear and cyberwar stratagems and consequences.

Professor Sanger points out that until the cyber age came along, America's two oceans symbolized our enduring national myth of invulnerability. The threat of nuclear attack preoccupied us during the Cold War, but generally, the United States has seemed confidently assured that it could take out dictators, conduct drone strikes on terrorists, and blow up missile bases in faraway lands with little fear of retaliation.

After some terrifying close calls, notably the Cuban Missile Crisis in 1962, we found an uneasy balance of power to deter the worst threats with our primary adversaries through aptly labeled Mutually Assured Destruction (MAD). Sanger notes that this common brutal understanding has worked thus far because the cost of failure on both sides is too high.

MAD stand-off nuclear threat deterrence between the United States and the Soviet Union seemed effective not only because each knew the other possessed world-destroying power, but also because each had confidence in the well-proven efficacy of its weapons systems.

As Sanger notes, in his book *The Perfect Weapon,* the past few years have witnessed time and again that cyber weapons can undermine that confidence. For example, and as discussed later, just as The North Koreans witnessed some of their missile launch tests go "mysteriously" awry, officials inside the Pentagon have good reason to fear that a similarly unanticipated failure might one-day sabotage a critical American defensive strike.

Or imagine an alarmingly plausible potential for skilled cyber hackers to compromise defensive American and/or allied early warning systems during an in-progress missile attack, just as Israel has likely accomplished against Russian-supplied air defense radar systems in bombing a Syrian nuclear site.

Alternatively, contemplate an opposite scenario where cyber hackers inject fake imagery into American or allied partner early warning systems indicating "false flag" evidence of an incoming missile attack to transfer attribution to a different adversary. Then further imagine that this might prompt the U.S. to respond with a retaliatory strike upon the wrong threat source.

Sanger believes that in the cyber age, not since MAD have opposing powers retained that strategic deterrence balance. Moreover, they probably never will because cyber weapons are entirely different from nuclear arms. He warns that although their effects have so far remained relatively modest, "to assume that will continue to be true is to assume we understand the destructive power of the technology we have unleashed and that we can manage it. History suggests that is a risky bet." [v]

Early cyber events to date present potential previews of

ominously impactful coming attractions. The joint U.S. and Israeli cyber sabotage of nuclear centrifuges in Iran also demonstrate technical capacities for other nations to shut down unambiguously vulnerable American power grids, energy pipelines, and critical communications networks.

Adversarial cyberwarfare proclivities and an eager willingness to apply them are broadly evident. For example:

- Russian attacks on Ukrainian power transfer and Internet communication networks spread to infect global banking systems and major corporate enterprises in many countries...including ours.

- Former U.S. Director of National Intelligence Dan Coats publicly stated that the Russian government had penetrated the control systems of some U.S. electric power companies, that we are in a period similar to the months before 9/11, and that "the warning lights are blinking red."

- Chinese government hackers are known to have penetrated thousands of communications networks in the United States and tens of thousands around the world including numerous national embassies. U.S. intelligence officials have reported that China operatives have also inserted prospectively destructive malware into the U.S. power grid.

- Iranian cyberattacks that froze financial networks of the Bank of America and Chase, fried computers at the Sands Casino demonstrate that their capacity for the ever-greater U.S. and global intimidation leverage.

- North Korean hacks into the Sony Corporation, the Bangladesh Central Bank, and even U.S. and South Korean government websites are but sample harbingers of far the greater damage that even a small hermit rogue nation can wreak upon economically and technologically advanced powers.

Whereas the USA, China and Russia were the first to invest

significantly in building cyber warfare capabilities, several intelligence studies claim that more than 140 countries are now believed to be developing cost-effective but effective cyber weapons.[vi]

In recent years, the development and use of these technological tools for offensive military and disruptive civilian operations have increased significantly. As a consequence, increasingly vast amounts of information are being stolen, and time-delayed kinetic attacks are being installed in critical infrastructures.

For all its countless benefits, this recent information and computational revolution have concomitantly spawned unfathomably terrifying cyberwar threats which might be comparable to World War II attempts to comprehend game-changing defense implications of the atomic bomb.

As Henry Kissinger wrote in his 1957 book *Nuclear Weapons and Foreign Policy*:

> *A revolution cannot be mastered until it is understood. The temptation is always to seek to integrate it into familiar doctrine; to deny a revolution is taking place.*

It was time, he said, to "attempt an assessment of the technological revolution which we have witnessed in the past decade" to better understand how that revolution affected everything we once thought we understood.[vii]

Attempts to understand the new cyberwar era should begin by recognizing that rather than occurring "out there" in military domains, it fundamentally targets and involves everyone with a cellphone, laptop, car GPS device, or even—in some cases—a TV that they mistakenly didn't know was listening to their bedroom conversations.

In other words, it involves everyone and everything at every time that is connected to the Internet of cyberspace. That includes military intelligence, government communications, the facial recognition cameras on the lamppost, your home security system, and the smart meter that monitors your energy use.

In this broad but realistic frame of reference, cyberwarfare combat targets include all computers, the "Internet of Things", and social media communication platforms that engage the entire "sphere of human thought." These cyberattacks can and do cause both physical "kinetic" destruction of equipment as well as "non-kinetic" assaults on personal and proprietary data including intellectual property, financial systems, and virtually everything in the realm of information, ideas and opinions.[viii]

Whether targeted on sabotage aimed at temporarily disrupting or permanently destroying critical infrastructures (such as communications, power grids, financial, and transportation systems) or espionage targeted on stealing private information (including security passwords, personal identity documents, and access to credit card and bank accounts), the consequences are incalculably vast.

As aptly summarized in the U.S. Department of Homeland Security's 2009 *Cyberspace Policy Review*, cyberspace touches practically everything and everyone:

> *[Cyberspace] provides a platform for innovation and prosperity and the means to improve general welfare around the globe. But with the broad reach of a loose and lightly regulated digital infrastructure, great risks threaten nations, private enterprises, and individual rights.*
>
> *The government has a responsibility to address these strategic vulnerabilities to ensure that the United States and its citizens, together with the larger community of nations, can realize the full potential of the information technology revolution.*
>
> *The architecture of the nation's digital infrastructure, based largely upon the Internet, is not secure nor resilient. Without major advances in the security of these systems or significant change in how they are constructed or operated, it is doubtful that the United States can protect itself from the growing threat of cybercrime and state-sponsored intrusions and operations.*
>
> *Our digital infrastructure has already suffered*

intrusions that have allowed criminals to steal hundreds of millions of dollars and nation-states and other entities to steal intellectual property and sensitive military information. Other intrusions threaten to damage portions of our critical infrastructure. These and other risks have the potential to undermine the nation's confidence in the information systems that underlie our economic and national security interests.

The federal government is not organized to address this growing problem effectively now or in the future. Responsibilities for cybersecurity are distributed across a wide array of federal departments and agencies, many with overlapping authorities, and none with sufficient decision authority to direct actions that consistently deal with often conflicting issues.

The government needs to integrate competing interests to derive a holistic vision and plan to address the cybersecurity-related issues confronting the United States. The nation needs to develop the policies, processes, people, and technology required to mitigate cybersecurity-related risks.

Information and communications networks are largely owned and operated by the private sector, both nationally and internationally. Thus, addressing network security issues requires a public-private partnership as well as international cooperation and norms. The United States needs a comprehensive framework to ensure coordinated response and recovery by the government, the private sector, and our allies to a significant incident or threat.

Just as the new Internet information revolution affords and promises unfathomably marvelous opportunities to advance beneficial human progress, let's also remember somber lessons tracing back to another technological information that emerged near the end of the Middle Ages.

The invention of the printing press spread both enlightenment and disorder in its wake. Tragically, the reformation process that followed also sparked a series of long wars that left Europe devastated and over eight million dead.

Perhaps more learned contemporary generations can figure out how to do better.

[i] *Cybersecurity and Cyberwar: What Everyone Needs to Know,* P.W. Singer and Alan Friedman, 2014, Oxford University Press.

[ii] *Cyber Warfare: The New Front,* Marie O'Neill Sciarrone, George W. Bush Institute, Spring, 2017.

[iii] *The Fifth Domain: Defending Our Country, Our Companies, and Ourselves in the Age of Cyber Threats,* Richard A Clarke and Robert K. Knake, 2019, New York, Penguin Press.

[iv] *Cyber War,* Richard Clarke and Robert K. Knake, 2010, HarperCollins.

[v] *The Perfect Weapon,* David E. Sanger, 2018, Crown Publishing Group, Penguin Random House LLC, New York.

[vi] *The 'cyberwar' era began long ago,* Ron Kelson, Pierluigi Paganini, Benjamin Gittins, and David Pace, June 25, 2012, The Malta Independent Online.

[vii] *Nuclear Weapons and Foreign Policy,* Henry Kissinger, 1957, Council on Foreign Relations, Published by Harper & Row, Inc. ISBN 978-0-393-00494-6.

[viii] *The 'cyber war' era began long ago,* Ron Kelson, Pierluigi Paganini, Benjamin Gittins, and David Pace, June 25, 2012, The Malta Independent Online.

Contents

Larry Bell

PROLOGUE

IMAGINE THAT EVERYTHING begins as a perfectly ordinary day in your life. It is mid-afternoon and you are casually engrossed in an online Skype conversation with a friend regarding places to visit on a trip you plan to take.

Suddenly, you just as you hear a loud boom the screen on your laptop goes blank. You soon learn that the sound resulted from the blowout of a nearby power transformer.

Hopefully, someone will think to call it in for repair.

Strangely, however, cellphone and landline services are down as well.

In this scenario, and unbeknownst to you, frenzied turmoil is unfolding in federal, state, county and local government offices throughout the country.

The National Coordinator for Security Infrastructure Protection and Counterterrorism, Security has just been notified that the White House Situation Room has received a "CRITIC" alert from NSA. She learns that the one-line message received before all secure communications went dead stated only: "large scale movement of several zero-day malware programs moving on the Internet in the US, affecting critical infrastructure."

The Director of the Defense Information Systems Agency has also just briefed the Secretary of Defense that the unclassified Department of Defense network known as the NIPRET is collapsing. Large-scale routers throughout the network are failing and constantly rebooting. Network traffic has essentially halted.

The Director of Homeland Security is attempting to establish secure contact with the Federal Emergency Management Agency (FEMA) following notifications that two of their regional offices in Philadelphia and Denton, Texas have reported large refinery fires and explosions. There are other reports of lethal clouds of chlorine gas being released from several chemical plants in New Jersey and Delaware. A major gas pipeline has exploded in a Virginia suburb.

Poison gas clouds are wafting toward Wilmington and Houston.

The U.S. Computer Emergency Response Team in Pittsburgh is being deluged with reports of mass system failures everywhere.

The FAA's National Air Traffic Control Center in Herndon, Virginia has experienced a total collapse of all systems. The alternate center in Leesburg is in a complete panic because it and several other regional centers cannot see what aircraft are aloft and are trying to manually identify and separate hundreds of aircraft.

Weather, navigation, and communications satellites are incapacitated. Even the U.S. military is struggling to reconnect communications between regionally and globally isolated units.

Brickyard, the Indianapolis Center has reported a midair collision of two 737s. Besides, the Federal Railroad Administration has been told of major freight derailments in Long Beach, Norfolk, Chicago, and Kansas City. Another derailment has occurred in a Washington Metro tunnel under the Potomac. Subways have also crashed in New York, Oakland, Washington, and Los Angeles.

Within a quarter of an hour, more than 150 metropolitan areas have been thrown into traffic snarls caused by signal failures.

The Secretary of the Treasury has been informed by the Chairman of the Fed that for some unknown reason their data centers and backup systems have lost all of their data. The entire U.S. financial system is frozen solid because terabytes of information have already been wiped out.

At this point, as lights briefly flicker and go out in the Situation Room, battery-operated emergency spotlights come on. Television Flat screens and computer monitors which went blank also return to life. Not so, however for most of the country that is in a full blackout mode.

Darkness everywhere, except where emergency generators provide isolated pockets of light and power. Soon, as batteries and fuel are exhausted, they will go out as well.

Hospitals are losing limited generator-supplied electricity to operate life-critical equipment. Meanwhile, primary power cannot be restored any time soon due to melted power transmission lines, damaged turbines at conventional power plants and lock-downs of nuclear plants.

Supplies of clean water are being depleted also. Metropolitan areas with high water towers atop high-rise buildings will have enough gravity flow to supply the most basic living needs for at best a few days. When that runs out, taps will go dry; toilets will no longer flush; emergency supplies of bottled water will become far too scarce for anything but drinking with no sources of replenishment.

Desperate food shortages are occurring due to distribution disruptions caused by rail system failures and jumbling of data at truck routing centers. Unable to withdraw cash from ATMs or bank branches, many Americans are being forced to go hungry, while others are looting stores. Police and emergency services engaged in rescuing people trapped in elevators have become overwhelmed. It seems that nearly everyone needs some kind of assistance.

Although most police are doing their level best to preserve

calm and maintain civic order, they, like everyone else, lack access to critical information and adequate intercommunications means to coordinate responses.

There is no way of predicting when power and fuel outages which have now affect tens of millions of people over several states. While current fall weather conditions are fortunately moderate, many of those regions anticipate a coming winter home heating crisis which will put countless lives at risk.

There are few opportunities for people to leave for warmer climes. Public transportation, including trains, buses, and airlines are encountering the same fuel shortages that private motorists are. Gasoline and diesel fuel can't be delivered, and gas stations can no longer operate fuel pumps because there is no power to enable them to do so.

Amounts of water, food, and fuel consumed by those isolated in large metropolitan areas stagger comprehension. So do the growing mountains of uncollected waste, including human biological material, which has to create an unthinkable sanitary and health crisis.

There is also a rapidly- growing crisis of conviction that anyone in government is really in charge of solving this unmitigated disaster. A contagion of panic and chaos ensues that leads to one inevitable conclusion.

There never was a plan to deal with well-known realities that such a threat existed.

Everyone, every family, every community of friends and loved ones, is ultimately now on their own.

The previous scenario unabashedly borrows (or steals) extensively from premonitions put forth by two writers have advanced such truly ominous warnings.

The first, Richard Clarke, served in the White House as National Coordinator for Security, Infrastructure Protection, and Counterterrorism during the Ronald Reagan, George H.W. Bush, George W. Bush, and Bill Clinton administrations. His projection of potential events is chronicled in the 2010 book *Cyber War,* co-authored with Robert K. Knake.[1]

The second source of these terrifying tidings is a 2015 book *Lights Out,* authored by former ABC news anchor and managing Nightline editor Ted Koppel.[2]

Koppel's dire premonition reportedly first arose in 2003 when a very large tree in Ohio had fallen on an electrical power transmission line. The event slowed down an overburdened control network resulting in cascading surges that tripped circuits providing electricity to 50 million people in eight states and two Canadian provinces. The blackout shut down everything that depended on the grid, including, for example, Cleveland's water system.

Fortunately, the outage lasted only a few hours for most people, and at most, a few days for others. And although a relief to many that it was later proven not to have resulted from a sinister cyberattack as many jittery people following the tragic 9/11 attacks two years earlier had suspected, it dramatically demonstrated how dependent we have all become upon power grid reliability.

A dozen years later, Koppel's book contemplated what could happen if a deliberate cyberattack on a U.S. power grid were to knock out large power transformers and generators and render them irreparable. With few available spares, it could well require many months to create new ones and restore services.

Ted Koppel's 2015 book struck lots of critics as being overly alarmist. The notion that malevolent hackers could shut down a grid—much less cause permanent physical damage to critical equipment—seemed more like the script of a science fiction thriller like the 2015 television series *Madam Secretary* where an American President retaliates against Russia by plunging Moscow into darkness in winter.

But could any of this scary hypothetical stuff happen in reality?

Ukraine: 2015

It was the day before Christmas Eve, 2015, when many citizens in western Ukraine were very much in a religious, good-will holiday spirit.

That was just when the lights went out.[3]

The timing had been ideal for saboteurs who hit Ukraine's power grid on that evening, and they weren't amateurs. They knew that the electric utility providers were operating with skeleton staffs, and for Vladimir Putin's army of patriotic hackers, Ukraine was a proven playground and testing haven for various sundry cyber spies, cyber vandals, and cyber burglars.

That timing also fits in well with Putin's military game plan. Less than two years earlier he had declared that Crimea was to be annexed to once again be part of Mother Russia. Is the support of his agenda, his "little green men"—soldiers who traded in their uniforms for camouflage in civilian clothing—were sowing chaos in the Russian-speaking southeastern Ukraine

It had been less than two years since Vladimir Putin had annexed Crimea and declared it would once again be part of Mother Russia. Putin's tanks and troops—who traded in their uniforms for civilian clothing and became known as the "little green men"—were busy doing sowing chaos in the Russian-speaking southeast of Ukraine. Their goal was to destabilize a new, pro-Western government in Kiev, the capital.

Cursors at Ukraine's Kyiv Oblenergo master control center suddenly began jumping across the screens as if guided by a hidden hand. Simultaneously, remote control hackers systematically disconnected circuits, deleted backup systems and shut down neighborhood substations.

Along with those disconnected circuits, the now-helpless operators' computer keyboards and mice had been disabled as well. The urgent situation was both bewildering and bizarre as if some paranormal powers had taken over their controls.

Putin's hackers also had other surprises for the controllers up their cyber sleeves.

The cheap malware program they had installed had been

designed to permanently wipe out those control systems. Then, adding insult to injury, the hackers disconnected the Kyiv Oblenergo control room's backup electrical system, leaving the operators in little to do but curse in the dark.

Whereas Ukraine had suffered previous cyberattacks, this one was far worse. Power and computers across the country shut down, causing everything that depended upon them to fail. ATM's were closed, radiation safety monitors at the old Chernobyl nuclear plant went offline, automotive fuel pumps couldn't operate.

News broadcast stations were interrupted; and when they came back on the air found communications blocked with what appeared to be ransomware notices on their frozen computer screens with a Broken-English message informing users that their hard drives had been encrypted. The notice stated: "Oops, your important files have been encrypted...perhaps you are busy looking to recover your files, but don't waste your time."

Stealthily disguised to appear as a financial shakedown, users were told that their computer data could be unlocked by paying a $300 ransom in hard-to-trace Bitcoin cryptocurrency. The masquerade was entirely a ruse to offer Russian government deniability cover.

Fortunately, the Ukraine cyberattack interrupted electricity for about 225,00 customers over a relatively brief period of a few hours. The impact would have been much worse had their grid not been controlled by old pre-computer-era metal switches in comparison with U.S. computer-dependent operations.[4]

The "NotPetya" virus used to attack Ukraine was ultimately traced back to a hacking organization known as "Fancy Bear" run by the Main Directorate of the General Staff of the Russian Federation's military, often called the GRU.[5]

Estonia: 2007

Russian hackers had been practicing and actively preparing for

11

cyberattack hits on communist regime targets for at least a decade before Ukraine. Such a plan was perpetrated in a relatively small coastal Baltic city of Tallinn on April 27, 2007, since known as "Bronze Night."

Tallinn, the population of about 400,000 people, had become the capital of an independent Estonia in 1989 following the Soviet Union disintegration into many component republics which dissociated themselves from Kremlin control. Prior to that time, Estonia had involuntarily been forced to become part of the USSR when the Red Army "liberated" the Baltic republic from the Nazis during World War II—which the Russians call the "Great Patriotic War."

The Bronze Night cyberattack on Tallinn is named after the removal of a giant, heroic bronze statue of a Russian soldier one of many erected in East European by the Communist Party to remind residents of sacrifices the USSR had made in "liberating" them. However, the free Estonians viewed the symbolism of those statues very differently—as painful reminders of five decades of oppressive by the Soviet Union.

In February 2007, the Tallinn legislature passed a Forbidden Structures Law requiring that anything denoting the occupation to be taken down, including the giant bronze soldier.

A heated dispute between Estonia and the USSR arose most particularly over the circumstance that moving the statues would defile graves of Russian soldiers interred beneath them. Relenting from Moscow pressure, the Estonian president vetoed the law.

Tensions between both ethnic factions escalated, catching both the police and the ghosts of the now-missing statue in the middle. Riots on April 27 soon provoked cyber-retaliation.

Servers supporting Estonia's most-often utilized webpages suddenly became so flooded with cyber access requests that many systems collapsed under the load and shut down. Since known as a "distributed denial of service" attack—or "DDoS" for short—this strategy has become a major weapon in international cyber arsenals.

DDoS generates a pre-programmed flood of Internet traffic which employs a network of remote-control "zombie" computers called a "botnet" to overwhelm and crash the Internet- connected systems. Those zombie computers can either be programmed to patiently await remote orders or can immediately begin to seek other computers to attack.

Botnets generally begin by taking down websites. Then next target Internet addresses that most people would not know such as servers that run parts of telephone networks, credit card verification systems, and Internet directories.[6]

Estonia, one of the most Internet-wired nations in the world, was particularly vulnerable to such attacks. Along with South Korea, the country ranks ahead of the United States in the extent of broadband penetration and broad public Internet utilization.

Accordingly, the Estonia DDoS was the largest which had ever occurred. The attack released several different botnets, each with tens of thousands of infected machines that had been sleeping and waiting for activation.

More than a million awakened computers became engaged in sending a flood of pings toward targeted Estonian servers. Hansapank, the nation's largest bank, was targeted, spreading to other on-line commerce and communications networks nationwide.

Unlike most previous one-time DDoS assaults, this one repeatedly hit one site for a few days, then moved on to attack another. Ultimately hundreds which were hit week after week were unable to recover.

Cybersecurity experts who rushed in from Europe and North America used trace-back techniques to follow the pings to specific zombie computers, then watched to see when the infected machines "phoned home" to their masters. The messages were traced back to higher-level controlling devices in Russia. The malware applied a software code determined to have been written as a decoy on Cyrillic-alphabet keyboards.[7]

The Russian government indignantly denied that it was engaged

13

in cyberwar activities against Estonia, although some officials admitted that perhaps some patriotic Russian hackers were responsible.

Following Bronze Night, the Russian security services had encouraged domestic media outlets to whip up patriotic sentiment against Estonia for desecrating heroic monuments and burial sites.

There is also a little secret that the most adept hackers in Russia, apart from those who are actual government employees, are usually in the service of organized crime which is profitably allowed to flourish. David Sanger reports that many close observers of Russia think that some senior government officials permit organized crime activity for a slice of the profits, or, as in the case of Estonia, for help with messy tasks.[8]

Did the Russian government security ministries engage in cyberattacks on Estonia?

Sanger proposes that perhaps this is not the right question. Instead: "Did they suggest the attacks, facilitate them, refuse to investigate or punish them? And in the end, does the distinction really matter when you are an Estonian unable to get your money out of a Hansapank ATM?"[9]

In any case, after Estonia brought the matter before the North Atlantic Council, NATO did respond by creating a new cyber defense center. It opened in 2008, just a few miles from the site where the giant soldier had originally stood.

Georgia: 2008

The Republic of Georgia, a region geographically slightly smaller than South Carolina with a population of about four million people, lies directly south of Russia along the black sea. Viewed within the Kremlin's sphere of influence as a Russian territory, the two nations have had a contentious relationship dating back nearly a century.

Georgia, a previous Russian territory, had declared independence in 1918 following the disintegration of the Russian Revolution. However, after the Russians finished fighting each other, the victorious Red Army once again took possession of Georgia, made it part of the Union of Soviet Socialist Republics, and installed a puppet regime.

Georgia once again declared independence in 1991 when the Soviet Union imploded in political turmoil.

Two years later, local populations in South Ossetia and Abkhazia, two territories still loyal to Moscow and supported by Russia, succeeded from Georgia to establish "independent" governments.

In early August 2008, Ossetian rebels or Russian agents (depending upon whom you believe), staged a series of missile raids on some Georgian villages. The Georgian army retaliated by bombing the Russian-aligned South Ossetian capital city followed by invading the region on August 7.

Georgian forces were then rapidly repelled by ejected from South Ossetia by Russian army forces. Russian aircraft bombed Georgian boundary sites to clear a "buffer zone" to isolate Georgia. Meanwhile, Abkhazia rebel groups succeeded in pushing out the remaining Georgians and expand the buffer.

Russian invasion foot soldiers were simultaneously joined by a small remote troop of cyber warriors whose primary mission was to prevent the isolated Georgians from knowing what was going on. They accomplished that goal by streaming DDoS attacks on Georgia media outlets and government websites along with access to outside CNN and BBC websites.[10]

In preparing the cyber battlefield—before physical attacks—hackers had previously conducted more limited DDoS hits on Georgian government websites. Included was the webserver of the president's site. In addition to defacing it, the hackers added pictures that compared the Georgian leader, Mikheil Saakashvili, to Adolph Hitler.

The cyberattacks rapidly picked up intensity and sophistication. Routers that connect Georgia to the Internet through Russia and Turkey were flooded with so many incoming DDoS that outbound

15

traffic was blocked.

Hackers also seized direct control of the rest of the routers supporting inbound traffic to Georgia. As a result, Georgia lost access both to outside news or information sources, and in addition, was denied means even to send an email out of the country.

As the Georgians attempted to accomplish "work-arounds" by shifting government websites to servers outside the country, the Russians countered every move. When they tried to block incoming DDoS traffic from Russia, the hackers rerouted their botnet attacks through servers in China, Turkey, Canada, and ironically, their former target victim, Estonia.

Georgia responded by transferring its president's government webpage to a server on Google's blogspot in California. As a further defense, the Georgian banking sector attempted to ride out the attacks by shutting down its servers altogether. The thinking was that it was better to suffer temporary financial online banking losses than to risk theft of critical data or damage to costly internal systems.

But the Russian hackers had prepared their work-around tricks for this occasion. Unable to penetrate Georgian banks, they deployed botnets that deliberately pretended to be a barrage of cyberattack traffic upon the international banking community originating *from* Georgia. Those attacks triggered automated Internet connections to the Georgian banking sector, paralyzing their operations. Credit card systems went down, soon followed by Georgia's mobile phone system.

The DDoS attack intensified and spread, ultimately enlisting barrages from six different botnets. The cyberassault infected and commanded computers from both unsuspecting and complicit volunteers. After downloading and installing software from numerous anti-Georgian websites, a volunteer could join the cyberwar by clicking on a button labeled "Start Flood."

Just as in Estonia, the Russian government incredulously claimed that the attack on Georgia was merely a populist activity that was beyond their control. Yet as pointed out by seasoned cyber expert Richard Clarke, any such large-scale activity in Russia, whether done by the government, organized crime, or citizens, is done with the approval of the intelligence apparatus and its bosses in the Kremlin.[11]

Clarke further notes that Russian nongovernmental hackers, including large criminal enterprises, are a substantial global force in cyberspace. Most are generally believed to be sanctioned by what was previously called the Sixteenth Directorate, a part of the infamous Soviet intelligence apparatus known as the KGB. Later it was called FAPSI, a Russian acronym for Federal Commission for Government Communications and Information.

FAPSI subsequently became what is now known as the Service of Special Communications and Information, an organization that runs perhaps the largest—and certainly one of the best—hacker schools in the world.[12]

America: Now

In the fall of 2017, The U.S. Department of Homeland Security began to quietly warn American power grid companies of a likelihood that potential adversary nations were attempting to penetrate controls over their systems. This information was already well known to some of them that had already been monitoring and observing such efforts for years.[13]

Other grid companies were more complacent, dismissing the penetration risk as small since their controls were not connected to the internet. And besides, they argued that it was solely the U.S. government's responsibility, not theirs, to protect them the citizenry from foreign attacks.[14]

By the summer of 2018, DHS's chief of industrial control systems analysis, Jonathan Homer, confirmed that Russia had successfully penetrated the U.S. power grid. Homer reported: "They [the Russian group, Dragonfly] have had access to the button, but they haven't pushed it."

Then-Director of National Intelligence and former Republican Senator Dan Coats described the Russian attacks and penetration of the U.S. electrical being so severe that figuratively, "the warning lights are blinking red." [15]

US government officials explained that the Russian hackers had "jumped the air gap", the disconnection between internal grid control system networks and the Internet which provides an open information highway everywhere. In reality, however, few of those companies had in reality isolated their controls from the Internet. Although many or most had segmented their internal networks by protective "firewalls", those precautions were seldom adequate to block penetrations by sophisticated hackers.

The Russians had dug into the grid infrastructure even deeper and wider, going after the companies that supply parts and do maintenance on the control side of the gap. Compromising those systems enabled attackers to tap into the log-in credentials of people with authorized access to the entire control network. This might allow hackers to remotely plug move into the systems that display the state of the grid on monitors inside the control rooms to send false instructions and readouts to thousands of field devices.

Russia was not the only threat. By 2019, heads of all seventeen U.S. intelligence agencies were receiving confirmations that in addition to Russia's ability to disrupt the U.S. power grid, China could sabotage both the U.S. power grid and natural gas pipeline system upon which the grid relies.

These are no longer theoretical threats. These are truly grim realities of today's new cyberworld. [16]

CYBERWEAPONS AND WARRIORS

BUT WERE THE previously referenced attacks upon Ukraine, Estonia, Georgia, and America, along with other (so far) non-lethal numerous which remain to be discussed, really acts of "war?"

And what if few or none directly lead to military or civilian deaths, what then? Where do government leaders—ours and others—draw the line? When should such attacks call for full-out military retaliation of a more conventional nature?

These are but a few key questions where no consensus answers appear to exist. There isn't any universally agreed definition of cyberwarfare, nor even whether it should be spelled as one word, two (cyber warfare), or hyphenated (cyber-warfare.)

For purposes of this book, the term "cyberwarfare" refers to a particular type of engagement—both offensive and defensive—with opposing government-state-connected actors that applies digital rather than physical kinetic weapons. Cybersecurity, in this regard, emphasizes digital blocking defenses against hackers and implants but also includes means and methods to identify and exact costly penalties upon those responsible.

Cyberwar is also presumed herein to signify a long-term conquest goal, a distinction that typically differentiates the "war" aspect from a singular event or limited series of cyber "attacks" which have narrower strategic and tactical objectives.

Whereas cyberattacks generally run lesser risks of escalating into "shooting" or even more destructive nuclear wars, their ultimate purposes may be to "prepare the battlefield" for just such options and eventualities. In cyberwar—when that day comes—the leaders of the target population may not know what or who was responsible until far too late.

It's reasonable to assume that there is little reason to hack into a power grid's controls, install a trapdoor to enable reentry at any time, and leave behind computer code that can be activated to hijack or destroy operations unless planning an option to do so. Yet planning and preparing for that option does not necessarily confirm an intent to implement action.

There appears to be growing public complacency regarding the idea that hostile foreign entities presently yield unthinkable disruptive power over critical American and global infrastructures.

There was a relatively little evident reaction, for example, to an April 2009 *Wall Street Journal* headline announcing that China had planted logic bombs in the U.S. grid. This fails to recognize that there is little distinction between the devastating effects those logic bombs could have on the power grid, compared to what little parcels of C4 explosives might do.[17]

Consequential differences between conventional warfare which applies explosive munitions and that which digitally explodes vital systems and services upon which our lives have increasingly come to depend are blurred distinctions. Add to this that cyberattacks have already demonstrated usefulness as force multipliers in conventional military assault operations.

Cyberattacks, for example, enable their sponsors and operators to cut off certain military units from higher command, and also to deny an opposing force access to intelligence about what is going on. An attendant danger is that cyberespionage and cybersabotage of a particular civilian or military command and control unit runs the risk of triggering a preemptive defensive counterattack that expands the battlefield far beyond the original target objectives.

Cyberespionage, the first stage of preparing any cyberattack, involves stealthy acts of covert hacking to obtain sensitive, proprietary or classified information from individuals and governments to gain military, political, or economic advantage. using illegal exploitation methods on the internet, networks, software and/or computers.

There are different types of cyber espionage depending on the scheme adopted to steal classified information that is not handled securely. The attacks could be conducted using malware to spy on victim systems, or by introducing/exploiting backdoors in software or hardware. This stage prepares the cyberspace battlefield for either and both conventional and cyberoffensives.

The typical primary aim of cybersabotage operations is to put means into place to destroy the target. Included are critical infrastructures such as power grids, communication networks, and economic systems such as an attack on the Israeli Tel Aviv stock exchange in 2012.[18]

Because of the interlocking nature of major global financial institutions, including individual banks, even a cyberattack on one nation's financial infrastructure could have a fast-moving ripple effect, undermining confidence globally.

Although cyberweapons will have far less physical destruction impact than nuclear weapons, their employment under certain circumstances can also be highly damaging and potentially trigger a broader war. In some cases, there might only be a few keystrokes of difference between penetrating a network to collect intelligence and the implantation deadly bombs of mass physical, economic and social destruction.

And while there has been no equivalent national or global

21

debate regarding threats posed by cyber versus nuclear weapons, the destructive threat and power of cyberweapons becomes more evident each year. The global inventory of these weapons appears to devices appears to be expanding exponentially, the ability of their creators to conceal their works and identities is becoming ever more effective, and the legion of hackers is growing both in numbers, sponsorship diversity and distribution, and technical sophistication.

Broad national defense strategies are also changing in this new cyber era. Whereas nuclear arms were designed solely for fighting and winning an overwhelming victory, cyberweapons, in contrast, come in many subtle shades, ranging from the highly destructive to the psychologically manipulative.

"Mutually Assured Destruction" (MAD) policies have been premised upon a logic that they can deter nuclear exchanges because both sides understand common vulnerability to utter destruction. Cyberweapons, and their expansive availability in hands of radical tyrants present an altered policy game-changing reality.

What most often comes to mind is the Cold War model of MAD as deterrence is that any attack would be met with an overwhelming counterstrike, one that would not only destroy the aggressor, but quite likely most life on the planet as well.

Cyberdefense generally operates on the premise that if the enemy can't get what they want by attacking, then they won't attack in the first place. This concept presupposes that the would-be attacker believes their identity is too well hidden to be discovered— the unknown "who" to retaliate against.

Unlike owners of military tanks and trajectories of missiles, cyber attackers don't leave correct return addresses unless they choose to.

Sometimes that lack of clarity is intentional in the confusing signals each opposing side sends to the other. For example, unlike the event of one side firing a missile, an intercept launch by the other makes it known that it and its source have been detected and successfully retaliated against.

In firing back with malware, on the other hand, the effect is not always so evident. It may simply appear to the aggressor as resulting from their faulty systems failure.

As a result, the cat and mouse game of cyberwar employs different attack and defense arsenals. "Noisy" cyberweapons often serve as "false flag" decoys, while stealthy ones can serve either purpose.

Richard Clarke, a Cybersecurity advisor to four white House administrations, emphasizes that vulnerability to devastating cyberattacks increases with levels of reliance upon the globally wired-together Internet of things.

Clarke points out, for example, that the US, South Korea and Estonia present special risks due to expansive consumer broadband access which includes Internet-capable mobile devices linked to computer networks. Those individual devices and networks, in turn, control electric power, pipelines, airlines, railroads, distribution of consumer goods, banking, and contractor support of the military.[19]

Our U.S. forces, in addition to being more wired, are also more dependent than any likely adversary upon private sector support. Clarke warns that even if the U.S. military's networks were secure and reliable, those of its contractors, who often rely upon the public Internet, may not be.

Put more simply, if you are going to throw cyber rocks, you had better be sure that the house you live in has less glass than the other guy's, or that yours has bulletproof windows.

Clarke adds that while such broad-based attacks would diminish a nation's military capacity, some military capabilities will suffer less than similar collateral damage to civilian infrastructure. This is because the military is more likely to have backup power systems, stockpiled food, and emergency field hospitals.

The last U.S. "Shock and Awe" Iraq war campaign employed accurately targeted precision-guided munitions that left nearby structures were left standing. As will be later be discussed, many applications of cyberweapons are—and will continue to be—much less discriminate in their attacks.

Unbounded Cyberspace Battlefields

Cyberwarfare moves faster and crosses borders far faster and more easily than the most advanced intercontinental missiles and aircraft. Once initiated, immediate expanding cascading effects pose a high likelihood that other nations will be drawn in.

Confirmed cyber activities by Russia, China, North Korea, Iran, and yes, the United States has established a very dangerous pattern and precedent which can only accelerate and compound national security risks. And whereas more significant hacking was previously undertaken by non-state actors, individuals, or clubs, today's major attacks are usually the work of some nation's military.[20]

Nations are now regularly using their militaries not only to steal secrets, but also to damage, disrupt, and destroy sensitive systems. Intentionally or unintentionally, such operations can readily escalate into broader wars.

By any definition, all proven weaponry and necessary warrior preparations for broad-scale cyberwarfare already exist. The United States, along with dozens of other nations are fully capable of devastating any other modern society.

Cyberwars, also by any definition, will pose very different deterrent and defense challenges than any previously encountered. For example, they will occur at the speed of light, meaning that the time before an adversary's launch of attack and its effects will be barely measurable and traceable, creating great problems for crisis response decision-makers.

In the nuclear age, response speed was key to MAD. It was crucial to be able to detect the early stages of an attack and launch retaliatory missiles before the other side's first strike. In the cyber age, however, that first strike might play out in nanoseconds. There are many compelling reasons to delay a counterstrike, such as to gain confident attribution and to plan the most appropriate and effective response.

Moreover, cyberwar attacks have effectively already begun. In recognition of mutual hostilities, many nations are actively preparing the battlefield: hacking into each other's networks and infrastructures, laying in trapdoors and logic bombs. This ongoing nature of cyber war, the blurring of peace and war, adds a dangerous new dimension of instability.

Richard Clarke observes that the recent and current levels, pace, and scope of disruptive activity in cyberspace by the military units of several nations are unprecedented, dangerous, and unsustainable. He urges: "It cannot continue like this. Either we control and de-escalate tensions, or conditions will cease to have any resemblance to peacetime."[21]

Clarke explains that whether or not we refer to these individual disruptive activities as "cyberwar", the collective result amounts to the same thing. He concludes:

> But then there were commentators and critics who thought such predictions were hyperbolic. By now, however, it seems generally accepted that this kind of warfare can happen.[22]

P.W. Singer, Director of the 21st Century Security and Intelligence at the Brookings Institution and Allan Friedman, Research Director of the Center for Technology Innovation at the Brookings Institution, point out that there are two key differences—in addition to those previously discussed—between war in the cyber realm and other past modes of conflict.[23]

The first difference, they argue, is that the indirect cyber effects will have less long-term destructive impact. While cyberattacks that change satellite GPS codes or shut down an energy grid, for example, would be quite devastating, they would have nowhere near the destruction visited by explosive-filled bombs and incendiaries upon Dresden, much less the nuclear holocaust experienced in Hiroshima and Nagasaki.

Having said this, let's also recognize that individual cyberattacks and broader cyberwarfare tactics now amplify—rather than simply supersede—combined threats posed by cyber and conventional arsenals.

Second, authors Singer and Friedman urge U.S. to consider that the weapons and operations in cyberwar will be far less predictable than traditional means, leading to greater suspicion of them among military commanders. Whereas the blast radius of a kinetic bomb can be projected to exacting standards; not so with the radius of most malware.

Since cyberattacks typically rely on second—and even third-order effects that might result—this widening of impacts can cause unpredictable and colossal collateral damage.

Singer and Friedman cite such an example that occurred during the Iraq war. The U.S. military launched a cyberattack intended to take down an enemy computer network used to facilitate suicide bombings. In the process, the operation accidentally took down over 300 other servers in the wider Middle East, Europe, and the United States, opening up "a whole new can of cyberworms."

It has now become common practice for government military entities to use cyberweapons against civilian targets to achieve more specific missions. Such purposes have included neutralizing a petrochemical plant in Saudi Arabia, melting down a steel mill in Germany, paralyzing a city government's computer systems in Atlanta or Kiev, or threatening to manipulate the outcome of elections in the United States, France, or Germany.

Government-state cyberweapon attacks are almost always employed at levels just below thresholds that will likely lead to retaliation. Singer and Friedman note (as further discussed later), that North Korea paid little price for attacking the Sony Corporation or robbing central banks. Similarly, China ultimately paid no price for stealing the most private personal details of 21 million Americans.

According to Singer and Friedman, best 2018 estimates of

known state-on-state cyberattacks over the previous decade or so numbered upwards of two hundred. That figure described only those events which were publicly released.[24]

Far less is publicly known about numbers of cyberattacks launched directly against civilian targets which may or may not be linked to state actor perpetrators. Such assaults range from notifications arriving in the mail warning that someone—maybe criminals, maybe the Chinese—just grabbed our credit cards, Social Security Numbers, and medical histories, for the second or third time.

Sometimes we first learn about those unpublicized events in newspapers or television news reports rather from government agencies or our trusted corporate finance and credit account holders and managers.

And still, after more than a decade of hearings in Congress, Singer, and Friedman that "there is still little agreement on whether and when cyberstrikes constitute an act of war, an act of terrorism, mere espionage, or cyber-enabled vandalism."

Cybersecurity experts Richard Clarke and Robert Knake observe that the main difference between ordinary thieves and world-class hackers is that with the best cyber thieves, you never know you were a victim.[25]

Clarke and Knake report that the U.S. government does many penetrations of foreign networks every month without being caught. Accordingly, they then ask if our government isn't getting caught, what aren't they catching when trying to guard our own personal, corporate and national security interests?

The two *"Cyber War"* co-authors warn that really good hackers, including the best government teams from countries such as the U.S. and Russia, are seldom stumped when trying to penetrate a network whether or not it is connected in any way to the public Internet. Furthermore, those varsity teams don't leave any marks they were ever there. The only exception is when they want U.S. to know.

The Hackers' Universal Toolkit

Whether perpetrated by nation-state opponents, attributed to real or false flag political patriot and terrorist organizations, independent or coordinated cybercriminal organizations, freelancing individuals and hacker clubs, or any combination of these elements, cyberstrikes apply common attack strategies and weaponry.

Although both the perpetrators and their methods vary from incredibly sophisticated to utterly basic, many general techniques are parts of a standard hacker toolkit which can be drawn upon to serve different purposes.

Often, several of these tools techniques are applied in a series of coordinated stages such as the case of the Russian 2007 Distributed Denial of Service (DDoS) attack on Estonia. Other attacks involve inside agents with simply steal malware and other highly confidential data from government and private networks.

The McAfee security firm has conducted extensive investigations of a vast "malware zoo" of various types of malicious or malevolent software viruses designed to wreak havoc on Internet users.[26]

In 2012, McAfee had discovered a new specimen of such malware every fifteen minutes. By 2013, the firm was discovering one every single second.

Sophisticated forms of viruses are continually becoming more and more difficult to detect.

Traditional antivirus software relies on scanning all files on a system along with incoming traffic against matching telltale "signatures" in a "dictionary" of known malware. This classic approach runs into growing pain problems. As the number and types of attacks rapidly expands the definition files must frantically try to keep that accelerating pace without lagging in search time.

Making the cybersecurity protection challenge even worse, most of the old signatures cease to represent current threats as new ones simultaneously proliferate.

Although one McAfee study found that only 3.4 percent of signatures in common antivirus programs were needed to detect all the malware found in all incoming emails, prudence dictates that all others must nevertheless also be scanned. This is a necessary precaution just in case an attacker gets sneaky and returns to reuse old malware.[27]

Hacking defense appears to be becoming a losing cyber tag-team chase.

Another McAfee analysis found that, over eight days, standard antivirus scans detected only twelve new malware signatures out of well over one hundred thousand which a major antivirus vendor had added to a list of what signature clues to look for. They attributed this alarming condition more to the increasing effectiveness of camouflage techniques used by malware authors rather than ineptitude of antivirus companies.[28]

By necessity, cybersecurity malware detectors are becoming more sophisticated as well. In addition to screening for malicious code signatures, they also apply "heuristic" detections to identify suspicious codes based upon analytical analyses of special hacker code-writing behavior patterns.

Hackers of all stripes and types, both proactively and defensively, apply common malware deception and deflection strategies.

Everyone connected to the Internet must be aware of imposter trolls purporting to represent a legitimate organization warning them that a payment account is overdue... or more, fortunately, they are informed a previously unknown relative has bequeathed them an enormous inheritance which is being held for them in a Nigerian bank subject to release of personal financial account transfer information.

This most fundamental and ubiquitous hacking strategy known as "phishing" tricks email and website users to turn over secret passwords and other data that allows culprits to assess proprietary personal, corporate and corporate files. Once the malware is downloaded, the scammers who are now "in" can often do pretty much anything they wish with that information. They can transfer money, read confidential emails, and gain access to other files and networks that interconnect with the infected user's private contact network.

Sometimes the phishing email provides a link to a phony website with a URL that looks very similar to a well-known authentic one.

If the attack is unsophisticated and reuses malware known to antivirus companies, a simple matching of the payload to fingerprints of known malware can often block it. However, in sophisticated phishing attacks, the fake page may also actually log the user into the real website to minimize the chance of detection.

Even more sophisticated attacks begin with "spear phishing" that selectively targets those holding authorizations to access high-value data networks. This strategy typically involves sending an email to a specific and unwitting employee at a company containing malware attached inside a pdf or some other document. The email often provides a link to a website, where clicking on it will automatically unload the malware will automatically download the virus.

Spear phishing attacks are typically customized to target specific individuals. Here it's like a distinction between receiving a personalized email that appears to be exactly like it's coming from a trusted friend versus one addressed to a broad field of recipients. Such specialized attacks require prior intelligence gathering to figure out how to trick that particular prime target.

"Firewalls"—a concept taken from a strategy of barriers built between buildings to prevent fires from spreading—offer the simplest form of network defense. These devices serve as filters that are intended only to permit valid network activities, and to block all others.

The next layer of a single computer and network defense employs "Intrusion detection systems" to detect known attack signatures and to identify invalid and anomalous behaviors.

Viruses of particular concern to critical networks are an insidious type of rapidly self-replicating malware called "worms". Once spread, they can cripple the network with an excessive overload of traffic, and sometimes deliver malicious payloads as well.

An example is the "Shamoon" virus that wiped out hard drives of more than 30,000 Saudi Aramco computers in 2012.

Worms and other malware exploit weaknesses in network intrusion detection systems and vulnerabilities in web browsers including various add-on components. Once downloaded, they can spread at exponential rates to cause drastic harm.

As discussed later, many worms that attacked Microsoft Windows in the late 1990s and early 2000s—which were known to but unreported by U.S. government security agencies—impacted global networks.

"Zero-day" malware attacks and exploits previously unknown code weaknesses that can give hackers access to or control over systems, but which have not yet been discovered and fixed by software companies. Such flaws are particularly prized because there will likely be no way to stop hackers exploiting them.

The zero-day term is taken from the notion that such attacks take place on the zeroth day of vulnerability awareness...thus before a patch to fix it can be implemented.

Knowledge about prospective zero-day exploits are valuable to both defenders and attackers. A common approach to gaining such information is to discover some way to trick a victim's computer into executing the attacker's commands rather than the those of the intended program's.

A thriving trade market exists for innovative zero-day malware that can slip through advanced security screens. Many nations are stockpiling these capabilities both for use in cyberespionage and for incorporation into elaborate cyberweapon exploits.

More and more countries are recognizing a need to develop and stockpile defensive and offensive cyberweapons just they do with conventional weapons. As a result, this will inevitably lead to patch more and more software security holes against zero-day attacks in endless cycles.

Compounding the challenge, zero-day worms and other stealthy sometimes come disguised as "patches" that users are fooled into installing into their security systems. Once installed they can allow an attacker to capture or destroy valuable data on network computers.

Malware can be spread over the Internet via "drive-by" attacks where the victim's only mistake is visiting the wrong website. Here, an attacker first compromises a web server, and then simply attempts to exploit vulnerabilities in any browser that requests files from that website.

Drive-by attackers often apply a so-called "watering hole" strategy that targets groups by going after websites used by specific companies. This approach draws upon lessons from smart lions. Rather than chase after their prey across the African savannah, they simply wait for all of them to come to a watering hole.

For example, a group out to steal secrets from a U.S. defense company indirectly targeted it by compromising the website of a popular aerospace magazine that many employees read.[29]

Worm-infected computers can be reprogrammed into "zombies" controlled within a "botnet" by outside actors. Botnets have become a common tool for nuisance Internet such as spam, activity, along with far more insidiously malicious distributed denial of service (DDoS) attacks witnessed in Estonia and Georgia.

Major DDoS attacks can release a botnet comprised of thousands or even millions of zombie computers to overwhelm computational bandwidth resources of a subsystem, such as web servers, that handle user connections to the Internet. This is analogous to having more and more people incessantly calling your phone. You first lose the ability to concentrate; then ultimately lose the ability to use your phone for other purposes.

Criminal gangs now routinely threaten to take targeted public and private websites offline with DDoS attacks unless they pay ransoms demanded for returns of interrupted services. These attacks range from small-scale hacks on individual bank accounts, global blackmail attempts against gambling websites before major sporting events like the World Cup and Super Bowl, and cyberassaults upon American cities.

Ransomware attacks are sometimes employed primarily to create political chaos. Supporters of the Syrian regime applied these DDoS tools to attach government critics and news organizations that covered the growing 2011 violence.

The US, UK and several other governments have blamed Russia for the NotPetya ransomware outbreak which caused havoc in mid-2017 which the White House described as 'the most destructive and costly cyberattack in history. Although most likely aimed at doing damage to Ukraine computer systems, it rapidly spread globally, causing many billions of dollars of damage.[30]

"Air gap" defenses against DDoS and other attacks attempt to physically isolate a computer or network from other unsecured networks, including the public Internet. This strategy is akin to one reportedly used by nuns to police Catholic school dances that stuffed balloons between teenagers dancing too closely to ensure nothing sneaky happened.

While this sounds simple in theory, and as history at and following those dances also likely demonstrated, it is a far more complex matter to achieve complete isolation in real practice.

Iranian nuclear weapons plant engineers learned this lesson the hard way.

Iran Nuclear Plant Gets a Virus Infection

In summer 2010, mysterious and disturbing occurrences began to confound technicians at Iran's Natanz nuclear facility. Nearly one thousand uranium enrichment centrifuges were suddenly breaking down for no apparent reason. These failures represented an estimated 30 percent of the plant's total enrichment output.

Before shutting down, those problematic centrifuges were exhibiting enormously strange behaviors. Sometimes they shook with damaging vibrations that caused them to grind to a stop. At other times, the same machines might spin rapidly out of control beyond design limits and explode.

Initially, assuming that their facility was just suffering from a series of random breakdowns, the engineers began replacing the broken centrifuges and failed parts with new ones. Those fixes didn't work because although the new ones checked out perfectly upon arrival, they soon began failing as well.

The beleaguered Iranian workers under the impression that they couldn't seem to do anything right, called in help from a Belarus computer security specialist to check what they began to suspect might involve control system sabotage.

As the security firm discovered, those suspicions were correct.[31]

It was later discovered that the Natanz plant cyberattack shut the centrifuges down in a particularly devious manner. An implanted "Stuxnet" worm had incorporated a series of very advanced subroutines. One, subsequently known as a "man in the middle," caused tiny adjustments in pressure inside the centrifuges. Another manipulated the speed of the centrifuges' spinning rotors to alternately slow down and spin up to damage the rotors.[32]

The insidiously deceptive design of the Stuxnet malware not only corrupted the operations of those centrifuges but also elaborately led the technicians to believe that the causes were somehow their fault.

The notion that some sort of virus might be implicated seemed entirely out of the question. After all, their control networks were isolated from outside the Internet by an air gap.

And besides, previous viruses had always had apparent effects upon a computer software. They didn't damage or destroy operational equipment hardware.[33]

The Stuxnet worm (as it was later called) proved to be truly innovative and complex.

As reported by P.W. Singer and Allan Friedman in their book *Cybersecurity and Cyberwar: What Everyone Needs to Know,* Ralph Langner and his German team of private security experts became fascinated by their discovery of malware codes with many of the same telltale Stuxnet-like characteristics that were infecting numerous large-scale industrial control systems not only in Iran but also in India, the US, and other countries.

Altogether, the more they dissected it, this new cyberworm of unknown origin possessed the most marvelously complex combination of malware code components they had ever seen.

For authentication, the malware utilized digital signatures with private keys of two certificates stolen from separate well-known companies. That package also contained new and old segments including at least four previously unreleased zero-day codes along with some that attacked vulnerabilities of all Windows operating systems dating back to the decade-old Windows 96 edition.

Stuxnet was designed to slip past Windows' defenses using the equivalent of a stolen passport gain access to and communicate with hardware devices connected with the target's operating control system. To accomplish this, the creators had opted to install a common "device driver" tool that allows trusted hardware to interact with that system. The Stuxnet worm drivers incorporated secret signing keys stolen from two real companies in Taiwan.

Most remarkably, Stuxnet incorporated an unprecedented number of four highly-prized zero-days all at once, something ordinary hackers would never choose to do. This alone offered evidence that whoever created Stuxnet had access to enormous resources. Besides, the style of attack using incredibly powerful, well-protected, and valuable stolen signing keys were also very costly and rare.

Even more interesting to Langner's group, rather than being broadly infectious, Stuxnet's sponsors wanted to be certain to achieve penetration of a very particular target. Unlike the goal of past worms, the code only allowed each infected computer to pass the malware to no more than three others.

Stuxnet also incorporated still another bewildering feature. It came with a self-destruct mechanism that caused it to erase itself in 2012. After doing its intended job, whoever designed it didn't want it lingering in the wild forever.

The malware's DNA revealed that Stuxnet was not going after computers—or even Windows software in general. Instead, it only targeted a specific type of program used in Siemens' WinCC/PCS 7 SCADA (short for "supervisory control and data acquisition") software. Otherwise, if this software wasn't present, the worm had built-in controls to become inert.[34]

Upon further analysis, Langner narrowed down the malware target not just any Siemens SCADA systems, but much more specifically to those tied together within a "cascade" of nuclear centrifuges of a certain size and number (984). Those requirements happened—not coincidentally—to exactly describe the setup at an Natanz site suspected of supporting an illicit Iranian nuclear weapons program.

Langner's group wasn't the only organization that was fascinated by remarkable Stuxnet features. David Sanger credits the worm's discovery to "cyber sleuths" Liam O'Murchu and Eric Chien of Symantec who ran it through filters, compared it to other malware, and mapped how it worked.[35]

Stuxnet was twenty times the size of the average piece of code. It became clear that something that large and complex containing almost no bugs couldn't be the work of an individual hacker or even a team of hobbyists.

It was later revealed that the delivery mechanism by-passed the imagined air gap by spreading expansively but undetected through the Internet to precisely target a particular configuration of centrifuges, and then worm its way to vital control systems through the Iranian nuclear scientists; laptops and memory sticks.

And while Stuxnet indeed comprised a very complex infection package that included novel and previously unknown zero-day malware codes, its most remarkable feature was the way these diverse and complicated attack elements all seamlessly worked together.

No doubt whatsoever, Stuxnet wasn't designed by a single group of very smart hackers. It could only then be achieved by a very major national or multinational effort costing millions of research and demonstration dollars.

We know now that the elaborate Stuxnet attack on those Iranian nuclear centrifuges—subsequently dubbed "Olympic Games"—was indeed a joint development of American and Israeli intelligence forces. They had even gone to the considerable expense of testing the malware on an expensive dummy set of centrifuges built just for the effort.

As reported by David Sanger, the CIA and the Israelis endeavored to slip the code in on USB keys, among other techniques, with the help of both unwitting and witting Iranian engineers.

With some hitches, the plan worked reasonably well for several years. Mystified and spooked regarding why some of their centrifuges were speeding up or slowing down and ultimately destroying themselves, the Iranians pulled other centrifuges out of operation before those met the same fate.[36]

Nevertheless, while initially devastating and dramatic, the overall physical damage caused by the Stuxnet worm wasn't a long-lasting setback. By most accounts, the Iranians lost about a thousand centrifuges, and out of fear of further destruction, the Iranian engineers took more offline. After the code leaked out, they put the pieces together.

After about a year the Natanz plant had not only recovered its original capacity, but ultimately about eighteen thousand new centrifuges—more than three times the number that had existed at the time of the attack.[37]

David Sanger concluded that the attack's more lasting effects were psychological, not physical:

> *When you look at a chart of Iran's production of enriched uranium, the Olympic Games were a blip, not a game-changer; a tactical victory, not a strategic one. But it created fear inside the Iranian nuclear establishment.*[38]

Although Stuxnet has come to be characterized by many as the world's first genuine cyberweapon in that it was designed to inflict physical damage, it will not be the last to do so.

Whereas it may have taken combined efforts and resources of two of the world's most advanced cyberforces to create Stuxnet, within weeks of its discovery an Egyptian blogger had posted an online how-to-guide on duplicating it. Yet despite such public postings, Sanger, Langner and other security experts lamented that even years later many major public infrastructure companies had still not plugged the vulnerabilities that Stuxnet attacked.[39]

During the summer of 2019, the Stuxnet virus got out in the wild and rapidly replicated itself in computer systems around the world. It was soon discovered in networks from Iran to India, and eventually even wound its way back to the United States.

Suddenly, everyone had a copy of it—the Iranians and the Russians, the Chinese and the North Koreans, and other hackers around the globe.[40]

Herding Cyber Copycats

Stuxnet copycat malware soon followed the release of its recipes. Some, such as Duqu which was discovered by the Budapest University of Technology and Economics in Hungary in September 2011, for example, used a very similar Microsoft Windows-exploiting code.

Duqu was named from the prefix "-DQ" it gives to the names

of files it creates. Its source broadly attributed to Unit 8200 (Unit eight—two hundred) of the Israeli Intelligence Corps.

Whereas many took Duqu as "son of Stuxnet"; merely a next-generation version designed by the same team. Key differences, however, indicate that it was more a case of inspiration than evolution. The malware was partially written in a previously unknown high-level programming language.

Ralph Langner described this new kind of proliferation trend in Singer and Friedman's book *"Cybersecurity and Cyberwar: What Everyone Needs to Know"*:

> *Son of Stuxnet is a misnomer. What's worrying are the concepts that Stuxnet gives hackers in reality. The big problem we have right now is that Stuxnet has enabled hundreds of wannabe attackers to do essentially the same thing.*

Langner adds:

> *Before, a Stuxnet-type attack could have been created by maybe five people. Now it's more like 500 who could do this. The skillset that's out there right now, and the level required to make this kind of thing, has dropped considerably simply because you can copy so much from Stuxnet.*[41]

Olympic Games and its Stuxnet development opened a door to a new dimension of warfare that remains as incomprehensible to U.S. today as thermonuclear weapons were to military strategists following World War II. And in both cases, there can be no backpedaling to return.

Retired four-star general Michael Hayden, who had been central to the early days of America's experimentation with cyberweapons, said that the Stuxnet code had "the wiff of August

1945 about it—a reference to the dawn of a new military era following atomic bomb explosions over Hiroshima and Nagasaki.

General Hayden had served as a former NSA Director, Principal Deputy Director of National Intelligence, and CIA Director during the beginning days of the Olympic Games. Although his security clearances wouldn't allow him to acknowledge the American involvement in Stuxnet, he left no doubt about the magnitude of its importance:

I do know this," Hayden concluded. "If we go out and do something, most of the rest of the world now feels that this is a new standard, and it's something that they now feel legitimated to do as well.[42]

A big proliferation difference between cyberweapons—which like Stuxnet exploit zero-day vulnerabilities—versus conventional bombs and missiles—is that they can rapidly and inexpensively be analyzed and repurposed by the adversary they were first used against.

Singer and Friedman point out that while building Stuxnet the first time which may have required an advanced team that was the cyber equivalent to the Manhattan Project, once used, the weaponry produced very different fallouts. it wasn't the same as if the U.S. only dropped a revolutionary new kind of bomb on Hiroshima. Besides, America kindly dropped leaflets with the design plan so that anyone else could build it too, and with no nuclear reactor required.[43]

One good example of this is a relatively unsophisticated May 2017 "WannaCry" ransomware attack which targeted the South Korean government in Seoul. The cyberattack (now known as "Operation Troy") broadly attributed to a North Korea hacker group known as "Shadow Brokers" used zero-day vulnerabilities developed for espionage by the U.S. National Security Agency.

Exactly how NSA's malware tool was acquired by the Shadow Brokers remains unclear. Nevertheless, after it became leaked it online, other ransomware writers incorporated it into their software, making it vastly more pervasive.

A booming and lucrative underground black market exists for creating and distributing malware, one where transnational criminal groups along with government entities buy and sell specialized cyber capabilities. This combination of elements fuels a new type of cyber arms race that no longer requires large-scale human, financial or physical resources once owned and controlled only by global superpowers.

Notwithstanding certainty that major cyberpowers including the US, Israel, Russia, and China will continue to break new ground in sustained cyberweapon advancements, lesser cyberforces and lower-end copycat actors are inevitably expanding global cyberspace threats.

As Joseph Nye, a former Pentagon official and dean of the Harvard Kennedy School observes that this configuration of power combines something old and something new: "Governments are still top dogs on the Internet, but smaller dogs [also] bite." [44]

AMERICA'S ALLIES AND ADVERSARIES

WHICH COUNTRIES ARE preparing for cyberwar? Although there are reportedly more than 30 nations that are developing government hacking programs of one sort or another. U.S. intelligence officials have suggested that those with emerging cyberwar capabilities include Taiwan, Iran, Australia, South Korea, India, Pakistan, and several NATO states such as France.[45]

Several nations, including the United States, presently or will soon possess capabilities to launch highly advanced and destructive cyberattacks including major disruptions of critical infrastructure networks.

But when the same kind of implants were discovered in American systems, the U.S. was outraged—understandably—and assumed the worst.

This, of course, is exactly what we were doing to Iran.

It is no secret whatsoever that Russia and China have probed U.S. networks for data that might be useful to "prepare the battlefield" for just such attacks. Assuming the worst motives—firm intentions to use them—top U.S. security officials expressed great concern.

Admiral Rogers, director of NSA and head of the U.S. Cyber Command until spring of 2018 rhetorically said:

> *We have seen nation-states spending a lot of time and a lot of effort to try to gain access to the power structure within the United States, to other critical infrastructure, and you have to ask yourself why.*
>
> *It's because in my mind they are doing this with a purpose, doing this as a way to generate options and capabilities for themselves should they decide that they want to potentially do something.*[46]

America: Zeus Cyber Command's Project Nitro

At the same time American security authorities worried—for good reasons—about the vulnerabilities of the U.S. electric grid, the Obama administration authorized a study of ways we could accomplish the same thing in Iran.

As reported by the *New York Times* national security correspondent David Sanger, also a national security policy instructor at Harvard's Kennedy School of Government, the US-Israeli Olympic Games collaboration to shut down Iran's Natanz nuclear enrichment plant was part of a much larger, more comprehensive planning program.[47]

Termed "Nitro Zeus," the cyber program explored ways to prepare the battlefield for possible future U.S. and/or Israeli airstrikes against Iranian nuclear sites by targeting their power grid, communications systems, and even their Revolutionary Guard Corps' command-and-control network.

In addition to providing attack cover, the strategy would also create confusion, buy time to photographically record how much damage had been done, and if necessary, bomb them again. The hope was that Nitro Zeus would also cripple Iranian missile defenses and retaliatory capabilities to such an extent that they would not be able to strike back.

In mid-2009, Nitro Zeus was assigned to a newly-created multi-service United States Cyber Command.

The need for a U.S. Cyber Command first emerged from a growing Pentagon awareness of outer space as a new war-fighting domain that originated in the early 1980s. In response, a Unified Space Command was established in 1985.

In 2002, Space Command was folded into the Strategic Command (STRATCOM) which operates strategic nuclear forces along with newly assigned centralized responsibility for cyberwarfare programs.[48]

America's cyberwarfare activities had originally been centered in a War Center established under the U.S. Air Force soon after the Gulf War in 1991. In 1995, the DoD-funded National Defense University graduated its first class of officers trained to lead U.S. cyber war campaigns.

By 2008, those in the Pentagon not wearing blue uniforms became convinced that cyberwarfare responsibilities were too important to be assigned solely to the Air Force under its own command. At that time the Navy had also followed the Air Force in setting up its cyberwarfare unit.

Although the Navy's cyber activities initially focused primarily upon espionage, it soon became obvious that once an information network was penetrated, it could be taken down with only a few more keystrokes.

A fierce political struggle ensued between the Air Force, Navy, and other intelligence agencies for control of this important and prestigious new cyberwarfare realm. A compromise that assured that while Cyber Command would be created as a multi-service agency, it would remain subordinated to STRATCOM as a "sub-Unified Command".

The compromise that created Cyber Command

("CYBERCOM") provided that it would be headed by the NSA Director—a four-star general who would be upgraded from a former three-star status position.

CYBERCOM focuses on five objectives: treat cyberspace as an "operational domain" as the rest of the military does the ground, air, or sea (and now outer space as well); implement new security concepts to succeed there; partner with other agencies and private sector; build relationships with international partners; and develop new talent to spur new innovation in how the military might fight and win in this space.[49]

In support of these mission objectives, CYBERCOM was directed to create and lead three types of cyber forces: "cyber protection forces" that will defend the military's computer networks, regionally aligned "combat mission forces" that will support the mission of troops in the field, and "national mission forces"—such as the U.S. Department of Homeland Security—to aid in the protection of important infrastructure.

A coded cryptographic message incorporated above CYBERCOM's logo states its mission statement as "9ec4c12949a4f314f299058ce2b22a". Translated, it reads:

> *USCYBERCOM plans, coordinates, integrates, synchronizes and conducts activities to: direct the operations and defense of specified Department of Defense information networks; and prepare, and when directed, conduct full-spectrum military cyberspace operations in order to enable actions in all domains, ensure US/Allied freedom of action in cyberspace and deny the same to our adversaries.*[50]

In 2010, the Pentagon co-located CYBERCOM Headquarters along with its Internet service provider alongside NSA to Fort Meade, Maryland (known as "The Fort") to protect DoD networks shared with other military cyber entities.

The reorganization agreement authorized the Air Force, Navy, and Army to continue to operate separate cyber war units as subordinate entities to the U.S. CYBERCOM. This arrangement enabled their war-fighting military units to engage in cybercombat activities which were technically out-of-bounds to NSA which is a partially civilian agency.[51]

In the transformations to create USCYBERCOM, what had been the U.S. Air Force Cyber Command became the 24th Air Force. Headquartered at Lackland Air Force Base in Texas. Although it had no aircraft, the 24th took control of two existing "wings": the 688th Information Operations Wing (IOW), and the 67th Network Warfare Wing.

IOW was charged to act as the Air Force's "center of excellence" in cyber operations. The 67th Wing assumed day-to-day responsibility for defending Air Force networks, and also for attacking enemy networks.

To support USCYBERCOM, the U.S. Navy reactivated its 10th Fleet which had been disbanded shortly after victory over Germany in 1945. And just as the U.S. Air Force cyber forces had no aircraft, the 10th Fleet had no ships.

The 10th Fleet had originally operated as a small World War II naval intelligence organization that coordinated anti-submarine warfare in the Atlantic. Then, as now, it operated as a tech-savvy "phantom" force.

The Army's cyberwarriors drew primarily from the Network Enterprise Technology Command of the 9th Signal Command at Fort Huachuca, Arizona. Their network warfare (NetWar) units, which operate under the Army's Intelligence and Security Command, are also forward-deployed to support combat operations alongside traditional intelligence units throughout the world. Included, is assistance working closely with NSA to deliver intelligence to warfighters on the ground in Iraq and Afghanistan.

An Army Global Network Operations and Security Center (known as A-GNOSC) manages LandWarNet, a service operated within the broader Department of Defense's networks.

A previously secret-level (now partially declassified) *"Military Strategy for Cyber Operations"* report contains a cover letter signed by the Secretary of Defense which declares the USCYBERCOM goal is "to ensure U.S. military [has] strategic superiority in cyberspace." It states that "Such superiority is needed to guarantee "freedom of action" for the American military and to "deny the same to our adversaries." [52]

To obtain a superiority, the U.S. must attack, the strategy declares "Offensive capabilities in cyberspace [are needed] to gain and maintain the initiative."

The DoD document implicitly acknowledges an unlimited sovereignty reach, wherein "the lack of geopolitical boundaries…allows cyberspace operations to occur nearly anywhere."

Further, it does not put civilian targets off-limits. It notes that cyber protection is "tightly integrated into the operations of critical infrastructure and the conduct of commerce."

Reflecting an increased offensive priority, the Trump administration upgraded the USCYBERCOM status in 2018 to that of a full and independent unified combat command within the U.S. Department of Defense. This now puts U.S. Cyber Command on the same level as groups such as the U.S. Pacific Command and U.S. Central Command.

According to DoD, U.S. Cyber Command plans, coordinates, synchronizes and conducts activities to:

> *[Direct] the operations and defense of specified Department of Defense information networks and: prepare to, and when directed, conduct full-spectrum military cyberspace operations to enable actions in all domains, ensure the US/Allied freedom of action in cyberspace and deny the same to our adversaries.*

The elevation of U.S. Cyber Command to independent unified combat status has changed the organization from conducting at most a few offensive operations each year, every one requiring a presidential authorization, to one more closely resembling its parent, the Strategic Command, which watches over America's nuclear forces. As such, they spend a great deal of time training, debating doctrine, establishing procedures for operations, and playing out scenarios.[53]

Meanwhile, as observed by David Sanger, NSA civilians working next door to U.S. Cyber Command remain far more risk-aversive. To them, years spent developing new cyber tools, learning the insides of Russian or North Korean or Iranian networks, and implanting their malware produces precious "implants" like prized bonsais, to be watered, nurtured, cared for, carefully hidden, and reserved for very vital occasions.

Accordingly, Nitro Zeus exposed many tensions between NSA—which possessed most of the talent needed to pull off the attack—and the military's newly created U.S. Cyber Command. The NSA invested huge resources into getting inside foreign systems, hiding its malware in hard-to-find corners, and checking in on it regularly.

By contrast, NSA tends to view the U.S. Cyber Command special operations military culture as one that likes to blow things up. They resisted exposures of their secret tools and methods through excessive exploits.

Nitro Zeus contemplated an aggressively ambitious plan; one that if enacted could well be considered an act of war. The U.S. had never before assembled a combined cyber and kinetic attack plan on this scale. In the process, Nitro Zeus littered Iran's networks with malware, placing implants in key strategic systems that could, later on, be used to inject destructive code or simply turn the networks off.[54]

Whereas Olympic Games (which was actually implemented), and Nitro Zeus (which was contemplated), both involved cyberweapons, they each did so for different strategic goals. Olympic Games was an NSA-led intelligence operation designed to help force Iran to the negotiating table. Nitro Zeus, on the other hand, was a military plan intended to unplug Tehran if diplomacy failed.[55]

David Sanger believes that the Iranians may have suspected that something like Nitro Zeus was in the works after Stuxnet along with the Olympics Games behind it became exposed.

But what they did know about American cyberattacks prompted Iran to begin building a cyber army of its own.

Yet as Sanger notes, even with their relatively limited cyber capability, the Iranians would expose a difficult truth about cyber conflict, one that Obama and other presidents must continue to grapple with.

The calculus of offense is inextricably wedded with that of defense. Whereby, defending the United States—with its sprawling financial systems, stock markets, utilities, and communications networks, all in private hands—remains next to impossible.

Israel: Stars of David Over the Euphrates

Just after midnight on September 6, 2007, many North Korean workers had previously left the construction site of a new building complex that was under construction in Kibar, eastern Syria. Suddenly there was a blinding flash. Yet observers didn't see the seven F-15 Eagle and F-16 Falcon Israeli Air Force strike formation with blue-and-white Star of David emblems emblazoned on their wings heading back home northward along the Euphrates towards Turkey, leaving that new building in fiery ruins.

The Israeli fighter jets had flown somehow flown deep into the Syrian interior and dropped several bombs, and as depicted in aerial photos, had leveled the complex. Throughout the entire times those aircraft were in Syrian airspace, the defense network never fired a shot.

Why hadn't those 1970s-vintage fighter-bombers been detected and targeted with missiles? After all, the Russians had provided the not-so-mysterious site with a modern air defense radar system that should have easily spotted their unstealthily designed steel and titanium airframes with sharp edges and corners.

Even though the Syrians had good reason to expect trouble, they didn't. Nothing had seemed unusual at all. They didn't even realize they were under attack until the bombs started going off.[56]

What appeared on those radar screens was what the Israeli cyber forces had put there, and even more important, what they didn't put there. No targets were calling attention from Syrian ground-based anti-aircraft controllers.[57]

Initially, Syrian President Assad asserted that what had been destroyed was an "empty building". However, a different story slowly emerged in the American and British media.

In April 2008, the CIA somehow obtained and publicly released a video showing clandestine imagery taken from inside the facility before it was bombed which left little doubt that the site had housed a North Korean-designed plutonium processing operation, a key development step in assembling nuclear bombs.[58]

The Syrians tried to clean up evidence of the facilities' true purpose. By the time the UN International Atomic Energy Agency (IAEA) sent inspectors to the site seven months later, it had been plowed and raked of any signs of debris or construction materials.

The plastic ziplock baggies of soil samples those inspectors brought back to their headquarters on an island in the Danube near Vienna told a different story than the Syrian government version. The surface scrapings unmistakably contained "man-made" radioactive properties.

Behind all of this mystery, however, was another intrigue. How could that defense system have been blinded after Syria had spent billions of dollars purchasing it from the Russians? This was certainly very bad publicity for its efficacy at a time when Moscow was in the process of selling the same air defense radar and missile technology from Moscow.

"Operation Orchard" or "Operation Outside the Box" as the cyber caper variously later came to be known, began in 2006 when a senior official in the Syrian government carelessly left his laptop computer in his hotel room while visiting London. Soon after he went out, agents from Mossad, the Israeli intelligence agency who had been flown in for the special occasion, snuck into his room and left him with more than he arrived with. They installed a Trojan horse trap door—a computer code containing lines of coded malware—into the laptop which enabled outsiders to monitor his communications.[59]

This trap door did even more than simply snoop. It provided a secret electronic access point that later allowed outside hackers to bypass the Syrian military network including its air defense intrusion-detection system and firewall; slip through its data encryptions; insert its own data streams; take full control of all administrator program rights and privileges; and substitute false images of what was really happening as the Israeli jets flew back and forth across their border.

As if that wasn't bad enough for the Syrians, the laptop computer files uncovered by the Israelis showed something perhaps equally consequential. One photo in particular that caught Israeli attention showed an Asian man in a blue tracksuit standing next to an Arab in the middle of the Syrian Desert. The Mossad identified the two men as Chon Chibu, a leader of the North Korean nuclear program, and Ibrahim Othman, director of the Syrian Atomic Energy Commission.[60]

The trick used by the Israeli's to penetrate and take control of the Syrian air defense network was similar to—and probably borrowed in part from—a playbook of programs developed by the U.S. in connection once (but no longer) top-secret Air Force cyberattack system code-named "Senior Suter" within a broader program called "Big Safari."

Senior Sutter was named after Colonel Richard "Moody" Sutter who created the USAF's Red Flag training program at the Nellis Air Force Base which focused on implementing electronic data warfare air defense systems. The USAF "senior" section of these projects dealt with electronic warfare in general. The Sutter program first emerged from its redacted status in 2002.[61]

Our Israeli friends have learned a thing or two from the programs the U.S. had been working on over at least three decades. Back in 1990, early U.S. cyber warriors joined together with Special Operations commandos to figure out how they could take out the extensive Iraqi air defense radar and missile network just before the initial waves of U.S. and allied aircraft came screeching in toward Baghdad.

Thirteen years later, well before initial waves of American fighter-bombers swept over Iraq, their military was informed that their "closed-loop" private, secure network had been compromised. The reason they knew this is because we told them so.

Just before that second war began, U.S. Central Command (CENTCOM) sent emails to thousands of Iraqi military officers. Although the actual content remains publicly unclear, they were likely advised to either walk away from their forces and

Thousands of Iraqi military officers received emails from U.S. Central Command (CENTCOM) on the Iraqi Defense Military email system just before the war started. Although the content remains publicly unknown, it likely advised them to walk away from their forces and go home or be killed.

Assuming that this was the message, many of those offers were taken seriously. Many Iraqi units parked their tanks in neat rows outside their bases making it convenient for U.S. aircraft to target them. Also, some Iraqi army commanders did send their troops on leave and went home, or at least they tried to.[62]

Russia: Moonlight Maze and a Fancy Bear

The Russian government and criminal enterprise hackers are major collaborating cyberspace forces generally believed to have been jointly sanctioned by a Federal Commission for Government Communications and Information (FAPSI).[63]

The organization was once known as the USSR's Sixteenth Directorate, a part of the infamous KGB intelligence apparatus. Their cyberwarfare operations are run by the Main Directorate of the General Staff of the Russian Federation's military, often called the "GRU." [64]

FAPSI later became the Service of Special Communications and Information which runs one of the largest and most advanced hacker schools in the world.[65]

Some of those hacking activities unambiguously target the U.S. power grid and other national security vulnerabilities.

In 1998, the FBI was called in to investigate seemingly bizarre intrusions that had been popping up in strange places connected to military and intelligence networks. Included were places where nuclear weapons are designed such as Los Alamos and Sandia National Laboratories, and quite conspicuously, the Colorado School of Mines.[66]

The discovery began when a computer operator at the School of Mines noticed some unaccountable nighttime network file activity. An investigation revealed that the hack attack was a very large one. Perpetrators, seemingly in Russia, had persistently stolen thousands of pages of unclassified yet technology-sensitive information over a period of two years.

It was a computer operator at the School of Mines who first discovered the hack after he saw some nighttime computer activity he could not explain. The attack turned out to be a very large one, and persistent, seemingly coming from Russia. The hackers had lurked in some of the systems for two years and had stolen thousands of pages of unclassified material concerning sensitive technologies.

Further analyses of the attack and its even broader and unprecedented snooping scope gave rise to intensified recognition within U.S. government intelligence and defense organizations of a looming cyber threat posed by Russia. The attack became named "Moonlight Maze." [67]

Russian Moonlight Maze technology hacking extended to a particular concentration of intrusions around the networks of Wright-Patterson Air Force base in Ohio, located on the site where the Wright brothers once tested many of their early planes.

Also in 1998, similar to at the Colorado School of Mines, a technician at the ATI-Corp, a specialty materials company, noticed early 3:00 AM Sunday morning computer activity connected with a network user account connected to Wright Patterson.

After confirming the account holder wasn't conducting business at that time, the technician raised an alarm to several CERTs (Computer Emergency Response Teams). The U.S. Air force was the first to respond. [68]

After confirming the attack, an investigation traced further connections to Wright Patterson from the University of South Carolina, Wright University and University of Cincinnati. In one instance, the attackers made a mistake in covering their origin by connecting from a machine traced to Moscow.

The Moonlight Maze investigation revealed that the Russian attackers made it a common practice to proxy activities through universities and small businesses which could provide valuable information with weak defenses. Having discovered this, investigators installed network equivalents of wire-taps at numerous compromised universities the attacks were moving through. This strategy enabled the cyber sleuths to monitor the culprits in real-time as they typed out their commands.

Released documents show that the investigators had also at least considered creating a "honeypot" of tempting information to lure attackers into a system designed to reveal more about their identity. This tactic had previously proven successful in exposing attacks involving a German hacker known as Markus Hess who had sold stolen technological information to the USSR during the 1980s. [69]

When confronted with questions regarding Russian Moonlight Maze involvement, Moscow blamed the intrusions on mischievous teenage hackers. U.S. evidence to the contrary ended all prospects for Russian government fact-finding cooperation.[70]

In 2008, just ahead of President Obama's election, Russian intruders reached higher into classified U.S. networks by boldly hacking into the Pentagon's SIPRNet "(Secret Internet Protocol Router Network) which connected the military, senior officials in the White House, and the intelligence agencies. Investigators were shocked in discovering how easily the Russians had gotten inside the top-secret network. They simply scattered malware-infected USB drives around the parking and public areas of a U.S. base in the Middle East. As intended, someone had picked one up, curiously put the drive in a laptop connected to SIPRNet, and presto—the Russians were inside.[71]

By the time the penetration was discovered, the bug had spread to all of the U.S. Central Command and beyond. It immediately began scooping up data, copying it, and sending it back to the Russians.

Following penetration discovery, the Pentagon wasting no time in response. The fix—called Operation Buckshot Yankee—was deployed by the Pentagon later that day. Then, as an extra precaution to keep a similar breach from happening again, they ordered that USB ports on DoD computers be sealed with superglue.

The vulnerability of sensitive and classified military and intelligence data revealed by Moonlight Maze gave NSA a new mission. In response, the agency zeroed in on protecting a vast set of targets; computer data stored around the world that was susceptible to a fast-growing cadre of ever-more sophisticated hackers.

As national security expert David Sanger points out, much of this information was and is not the kind of "data in transit" that NSA had spent decades intercepting: "Instead, it was locked away in computer complexes that governments, in their naiveté, had viewed as largely invulnerable."

Sanger observes that whereas NSA had previously focused its primary attention "intercepting electrons flying through phone lines and over satellites", the new priority called for capturing what the agency calls "data at rest." And getting that kind of data requires breaking into computer networks around the world.

As for Moonlight Maze, in many respects the Russian cyber campaign never really ended; but only morphed into new verities of ongoing cyberattacks.

The Russians are probably saving their best cyber weapons for when they need them, in a conflict in which NATO and the United States are involved. Moreover, they are learning many of their new cyber tricks from America.

The 2017 NotPetya virus attack on Ukraine to disrupt its financial systems were ultimately traced back to a GRU-connected organization known as "Fancy Bear."

How do we know this? It so happened that Israel's military intelligence Unit 8200 malware had been inserted inside the Russian Kaspersky Anti-Virus packages installed on computers around the world. The Israelis watched the Russian hackers as they—in turn—watched everything go down from their Moscow headquarters-based network.[72]

Our NSA would certainly have known all about this right away also. The Israelis would have informed them that in autumn of 2013 they had hacked into a GRU server used to store attack tools, a so-called staging server, that contained NSA hacking tools. One of those NSA tools, a "Sandworm" known as "EternalBlue" looked a lot like NotPetya.[73]

And how can we be certain that the Russians were the ones who hacked into NSA's crown jewel files? We know because in the summer of 2016, a Russian GRU group posing under a fictional alias of "Shadow Brokers" proudly posted them online for all the world to see. The hacked files all seemed to date back to 2013.[74]

In launching the NotPetya attack, the GRU had noticed was

that almost every company and government ministry in Ukraine used the same accounting software known as "M.E.Doc" from the Ukrainian software company—the Linkos Group. M.E.Doc program updates were regularly issued to licensed Linkos user networks with digitally-signed authentications.

The GRU had hacked into Linkos and succeeded in planting a NotPetya malware package into one of those M.E.Doc updates that exploited a vulnerability in the users' Microsoft server software. The worm included a password-hacking tool along with instructions to spread to any connected device on the network, wiping them of all software.

Whether or not the GRU intended for all global companies operating in the Ukraine to be hit, the impact was truly devastating. The attack destroyed an estimated ten percent of all devices in Ukraine, including some in every government ministry, more than twenty financial institutions, and at least four hospitals. It also spread over virtual private networks (VPNs) and rented corporate fiber connections from Ukrainian offices back to corporate headquarters in England, Denmark, the United States, and elsewhere.[75]

Microsoft charged that NSA had previously known about the existence of their server problem for five years, yet failed to inform them. Instead, NSA developed EternalBlue for a presumed purpose of exploiting Microsoft's vulnerabilities to penetrate foreign networks.

Upon belatedly becoming aware of their problem by the hackers, Microsoft quickly issued a patch to fix it. Nevertheless, as with all patches, not all users got the notification or immediately acted to install it.

Even despite the patch, the North Korean authors of a copycat WannaCry virus successfully exploited the same Microsoft vulnerability only two months late in May 2017.[76]

The Russian GRU repeatedly used it again also.

Richard Clarke and his coauthor Robert Knake draw two important lessons from Russia's NotPetya attack on Ukraine. The first is that nation-state military and intelligence organizations are already taking down major global companies such as Maersk and Merc with cyberattacks.

The second lesson is that their attacks can be targeted somewhat selectively. The other U.S. and global companies in Ukraine during the NotPetya attack including Hyatt Hotels, Abbot Laboratories, Boeing, Dow DuPont, Eli Lilly, Johnson & Johnson, Cargill, Pfizer, Delta Air Lines, and John Deer weren't significantly damaged.[77]

Top U.S. government officials were fully aware of activities by Russia, among others, to hack into vital U.S. infrastructure networks years before the Ukraine NotPetya attack.

American intelligence agencies had been warning that Russia was likely already inside the American electric grid since at least 201 when Russian malware is known as "Black Energy" was detected in software controlling U.S. turbines.[78]

Indeed, Russia's successful penetrations of U.S. highly classified Pentagon, State Department and White House networks along with vital energy grin and communications infrastructures represents a sustained espionage campaign dating back at least to its 2007 attack on Estonia and its 2008 attack on the nation of Georgia.

Even more directly, the GRU attacked the Ukrainian power grid in 2015, and again in 2016.

According to UK intelligence sources, the GRU also engaged in cyberattacks to interfere with investigations of a Russian assassination attempt in Bristol, England; the Russian doping of Olympic athletes; and the suspected Russian downing of Malaysia Airlines Flight 17. These operations were conducted under cover of the false flag name of "Cyber Caliphate", an apparent attempt to sound like an Arab terrorist group.[79]

Richard Clarke emphasizes that although many of the Russian attack tools can be traced back to U.S. cyber arsenals, we should not conclude that the GRU necessarily needs to steal most of them. He warns that they already have plenty of very good ones they have developed themselves:

> *Their motive in stealing and publicly releasing the U.S. cyber arsenal is to embarrass the United States, make it seem like America is the world's most problematic hacker, and allow nations (including our friends and allies) to go back and identify U.S. intelligence operations against them (thereby creating distrust among allies).*

And if not necessarily the most skillful of all global cyber powers, the GRU, along with its complicit non-state pretenders and surrogates, clearly ranks highly among those most reckless and indiscriminate.[80]

As discussed later, the 2015-2016 Russian penetrations of the Democratic National Committee (that actually required very little skill), represents a particularly famous example.

China: An UglyGorilla in Unit 61398

Viewed from the outside, the bland twelve-story building along the outskirts of Shanghai, a city of 24 million people, looks pretty much like any other boxy structure in that run-down Pudong neighborhood populated with massage parlors and noodle joints. A big difference, however, is that attempting to photograph it will bring a swift and unfriendly response from Chinese security forces assigned guard its occupants.

The building houses Unit 61398, formally the 2nd Bureau of the People's Liberation Army's General Staff Department's 3rd Department—China's premier cyber spying force.[81]

Workers inside the building appear quite ordinary also. Mostly males in their mid-twenties, they arrive at about eight-thirty a.m. Shanghai time, check sports scores, email girlfriends, and some occasionally watch porn.

Promptly at nine, they put personal interests aside and begin their daily routine, banging on their computer keyboards to hack into computer systems around the world until lunchtime rolls around.

They will then have another free hour to check out more sports updates, girlfriend chat, and porn.[82]

How can we possibly know this?

It's because by 2012, American intelligence agencies who had been watching Unit 61398 activities for several years, had succeeded in planting malware inside their network which enabled them to activate cameras on the hackers' own laptops. This allowed hackers of those hackers to watch them at their desks and monitor their keystrokes as they attempted (often successfully) to break into "secure" global computer networks.

Viewing those keystrokes enabled NSA operatives to track down the hackers' individual identities through privileged Facebook accounts which weren't accessible to ordinary Chinese citizens. One of those accounts linked to a particularly prolific hacker using the screen name "UglyGorilla."

As reported by Davis Sanger, NSA observers watched UglyGorilla and his fellow workers log in and steal blueprints and identification numbers from RSA, the American company best known for making the SecurID tokens that allow employees at military contractors and intelligence agencies to access their email and corporate networks. The hackers then used the data stolen to hack into Lockheed Martin's network.[83]

in 2013, 61398 was caught stealing employee passwords to break into the *New York Times'* computer networks. The *Times* got its revenge by publishing a series of articles exposing the once-secret unit had been targeting cyberattacks across 20 different industries and governments ranging from Coca-Cola, to the Pentagon, to the United Nations. The newspaper even published a front-page picture of its no-longer-secret headquarters on Datong Road in Shanghai.[84]

Unit 61398 is but one of a great number of cyber-espionage organizations that are directly or effectively subordinate to China's

People's Liberation Army (PLA). Known in cybersecurity circle also as the "Comment Crew" or "Shanghai Group" the unit is tasked with key responsibilities for gathering U.S. political, economic, and military-related intelligence.[85]

Another PLA Unit 61486, nicknamed by the NSA as "Putter Panda", is especially known for cyberattacks on U.S. infrastructure and defense systems, as well as European satellite and aerospace industries. Together, just Unit 61398 and Unit 61486 alone have penetrated thousands of networks in the United States, and tens of thousands around the world.

The US-China Economic Security Review Commission estimates that there are up to 250 groups of hackers in China that are sophisticated enough to pose an "Advanced Persistent Threat" (APT) to U.S. interests in cyberspace.[86]

Some cyber programs are co-located with engineering schools and technology firms. Unit 61539, for example, is next to Beijing University and the Central Communist Party School in the city's northwestern Jiaoziying suburbs.[87]

Unit 61539 is a key part is the Beijing North Computer Center (also known as the General Staff Department 418th Research Institute) which some believe to be the Chinese equivalent of the U.S. Cyber Command. Altogether, it has at least ten subdivisions involved in "the design and development of computer network defense, attack, and exploitation systems."[88]

Although universities afford prime cyber talent recruiting sources, they don't always keep secrets very well. Unit 61398 drew U.S. special public attention when it ran a remarkably public notice on the Zhejiang University website stating that "Unit 21398 of China's People's Liberation Army (located in Pudong District, Shanghai) seeks to recruit graduate students of Computer Science of Class 2003."[89]

P.W. Singer and Alan Friedman estimate that there are at least twelve additional PLA cyber training facilities located around the country. Many draw upon cyber expertise resident in its eight-million-strong people's militia, supplementing official forces with a "patriotic hacker" program.

Of special note is a unit located in Zhurihe that is permanently designated to serve as an "informationalized Blue Team." The unit simulates war game scenarios to assess how the U.S. military and its allies might use cyberspace in ways that provide strategic targets for Chinese units to counterattack.[90]

Beijing and its PLA have been rapidly building China's cyber espionage and warfare capabilities as top military priorities. Since the late 1990s, the government has invested heavily in all things a nation would do when preparing for proactive and reactive attacks. It has created military and citizen hacker groups; engaged in extensive cyber espionage, including U.S. computer software and hardware, and established cyber war military units, and laced U.S. infrastructure with logic bombs.

China's Communist Party first organized once-informal recreational hacker groups into more formal, controllable organizations. An important step in this direction took place in 2003 in the Guangzhou Military Region, where part of China's IT economy is clustered.

PLA military officers conducted a survey to identify "those with advanced degrees, who studied overseas, conducted major scientific research, [and] those considered computer network experts." [91]

Unpaid, volunteers (as much as that means in such an environment) units were organized operated out of computer labs and commercial firms rather than military bases. This provided the militia a combination of "politically reliability", educated staff, modern commercial infrastructure, and advanced software design capabilities.

Such unofficial units are now estimated to have well over 200,000 members.[92]

By 2003, China had established cyberwarfare units at their naval base on Hainan Island, a Third Technical Department of the

PLA, and a Lingshui Signals Intelligence Facility.

According to the Pentagon, these units are responsible for both offensive and defensive operations that design and apply unknown cyberweapons that that no defenses are prepared to stop.

A Chinese publication listed ten examples of such weapons and techniques:

- Planting information mines

- Conducting information reconnaissance

- Changing network data

- Releasing information bombs

- Dumping information garbage

- Disseminating propaganda

- Applying information deception

- Releasing clone information

- Organizing information defense

- Establishing network spy stations[93]

China has established two "network spy stations" with permission from the Castro government in Cuba. One monitors U.S. Internet traffic; the other monitors Department of Defense communications.[94]

A three-year-long series of Chinese cyberattacks on American computer systems dating back to 2003 since known as "Titan Rain" extracted between 10 and 20 terabytes of data off the Pentagon's unclassified network. Traced back to a server in Guangdong, China, the hackers also targeted the defense contractor Lockheed Martin, other military sites, and the World Bank.[95]

By 2007, Chinese cyber espionage expanded to widespread penetrations and massive data thefts from U.S. and European networks.

The Director of the British domestic intelligence service M15 accused Beijing's government of hacking into the computer of German Chancellor Angela Merkel. U.S. Secretary of Defense Robert Gates was also targeted.

The Chinese the operatives later copied information taken from U.S. Secretary of Commerce Carlos Gutirrez's laptop when he visited Beijing, then attempted to use that information to gain access to Commerce Department computers.[96]

Two years later, Canadian researchers uncovered a highly sophisticated computer program they dubbed "GhostNet." By the time they had discovered it, the malware had taken over an estimated 1,300 computers at several countries' embassies around the world.

Chinese hackers simultaneously conducted a series of intrusions later referred to as "Byzantine Hades" attacks against nongovernmental organizations working on Tibetan issues including the office of the exiled Dalai Lama. They did so by tracking communications from infected computers back to control servers that had previously gone after Tibetan targets during the 2009 Olympics in Beijing. While the origin of the operation was never confirmed, the servers utilized were all located on Hainan Island in China.[97]

GhostNet could remotely and secretly turn on a computer's camera and microphone without alerting the user and to export images and sound silently back to servers in China—perhaps not surprisingly just as NSA did in keeping watch over China's Unit 61398.[98]

The GhostNet operation ran for twenty-two months until first discovered in 2009. That same year, U.S. intelligence leaked to the media that Chinese hackers had penetrated the U.S. power grid and left tools that could be used to bring it down.[99]

According to Richard Clarke, a former government official told him of suspicions that the Chinese actually wanted U.S. to know—as a threat—that if we intervened in a Chinese conflict with Taiwan, the U.S. power grid would "likely" collapse: "They want to deter the United States from getting involved militarily within their sphere of influence."

If the Chinese truly planned to attack our electric grid, they would likely have more secretly planted logic bombs in ways that wouldn't have exposed vulnerabilities so that we wouldn't have noticed.

Richard Clarke observes that over several years during the mid-1990s, the Chinese talked very openly, for a Communist police state, about what they had learned from the Gulf War. They concluded from that experience that a strategy to defeat the U.S. with overwhelming battle forces would no longer be successful.[100]

China then began to downsize its military and invest in new technologies that included "wang guohua" (networking) to deal with the "new battlefield of computers."

As one Chinese expert wrote in his military's daily paper, "the enemy country can receive a paralyzing blow through the Internet."[101]

A planning maxim dating back from the 500 BC teachings of the great warfare strategist Sun Tzu teaches that "a superior force that loses information dominance will be beaten, while an inferior one that seizes information dominance will be able to win."

Or as Mao Zedong advocated in the 1940s:

> *To achieve victory, we must as far as possible make the enemy blind and deaf by sealing his eyes and ears and drive his commanders to distraction by creating confusion in their minds.*

Washington Times national security columnist and Washington Free Beacon Senior Editor, Bill Gertz agrees with Richard Clarke that full Chinese commitment to military cyberwarfare is no secret.

In March 2014, three PLA officers laid the strategy out for

everyone to see in the *Guangzhou Military Region* newspaper story headline titled *Carry Forward the Thinking on People's War, Win Cyber Network War in the Future.* Their feature article explained that adapting "people's war" to cyberspace is the key to helping a weaker Chinese military defeat a stronger United States:

> *To wage people's war in the cyber network era, we cannot expect any readily available prophetic answer from any great man, nor can we copy experience and practices in a simplified manner.*
>
> *How to inherit and carry forward the thinking on people's war and how to engulf our enemy in a 'boundless ocean of people's war' are major mission-related topics that are worth our great attention and study [at the PLA Combat Development Center].*[162]

The PLA began its "boundless ocean of people's war" in the 1990s with large-scale cyber espionage attacks targeting foreign sources of secret science and technology information and innovations. By the end of the 1990s, the Chinese government strategists had created an "Integrated Network Electronic Warfare."[103]

These electronic warfare programs were, and continue to be, aimed at rapidly expediting China's global military and industrial sector prominence—and to accomplish such advancements at very little cost through theft.

Among at least nineteen confirmed and nine possible PLA cyber units, one identified as "3PLA", also known as the "Technical Reconnaissance Bureau," is the most aggressive by far. A 2018 intelligence report released by the Trump administration named one of China's leading spymasters, PLA Major General Xiaobei Liu, as its previous director.[104]

General Liu, the son of former Guangzhou Military Region deputy commander PLA Lieutenant General Liu Changyi, had been promoted to that position in 2011, most likely in recognition of his enormously successful cyber thefts of American technologies. He appeared in a 2013 PLA propaganda video called "Silent Contest," which described the United States as the main target of Chinese cyberattacks based on the country being the birthplace of the internet and having the ability to control its core resources.

Liu accused the U.S. of subverting Chinese Party rule through dominant public Internet influence, saying:

> *The U.S. took advantage of its absolute superiority of the internet and vigorously promoted network interventionism to reinforce ideological penetration, and it secretly supported hostile forces to create obstructions and conduct acts of sabotage.*

General Liu emphasized the importance of weaponizing the Internet to China's advantage:

> *The internet has become a new field and platform for ideological struggle. Accordingly, we must not lower our guard; [we] must take control of the commanding heights of the internet and maintain both the initiative and discourse power.*[105]

In 2015, 3PLA became the core unit of a new military "Strategic Support Force" and, in turn, the main component of a new "Cyber Corps." Headquartered in Beijing's Haidian District, the new corps is one of PLA's most secret units. Together with branch units in Shanghai, Qingdao, Sanya, Chengdu, and Guangzhou, it is staffed by as many as 100,000 hackers, language specialists, and analysts. A psychological warfare group called the "311 Base" specializes in waging public disinformation and influence campaigns.

A declassified report first disclosed and made public by the U.S. Trade Representative (USTR) Office reveals how General Liu "and other senior intelligence officials" directed 3PLA cyber-spying operations against American oil and gas companies during talks with officials from the state-owned China National Offshore Oil Corporation (CNOOC). The spying targeted several oil and gas companies that were working with cutting-edge shale gas technology to enable a Chinese energy giant to beat an American company in a deal.

The extent of PLA cyber spying against US, European, and Japanese industries and research facilities is without precedent in the history of espionage. Exabytes of data stolen from universities, industrial labs, and government facilities reveal secrets behind everything from pharmaceutical formulas—to bioengineering designs—to nanotechnology—to weapons systems—to everyday industrial products.[106]

The costs to targets of these thefts are staggering. The U.S. Trade Representative Office has reported that Chinese technology theft and unfair trade practices cost Americans between $225 billion and $600 billion in lost information annually.[107]

An internal July 23, 2008 email from a PLA officer led to a series of events that exposed China's hugely successful cyber campaign to transfer of American wealth. The communication referenced massive technology thefts ranging from extremely valuable government information to the pillaging of proprietary electronic data on some of the most strategic U.S. weapons systems.

All of these thefts covertly targeted a small of defense contractors including Boeing and Lockheed who built and maintained cutting-edge aircraft, warships, and other military hardware.[108]

The same PLA officer later sent an internal October 23, 2009 email containing a draft contract (signed the next day) for a "System for Unidirectional Secure Delivery of Files Over the Internet" from a known Chinese company that had advertised its ability to conduct computer network attack and defense and communications security. The firm was well known by U.S. intelligence agencies to have ties both to China's Ministry of State Security intelligence service and to 3PLA's hacking operation linked to every major Chinese cyberattack on English-language-based targets.[109]

Over the next five months, a contracted team of hackers operating in China began targeting specific employees with access to computer networks at the Boeing C-17 assembly plant in Long Beach, California with spear-phishing emails that were carefully crafted to masquerade as someone known to the recipient.

The cyberattacks paid off handsomely, netting the hackers 85,000 Boeing C-17 aircraft files between December 2009 and January 2010.[110]

An intercepted August 12, 2013 email from an operative to PLA higher-ups under a subject line "C-17" outlined the successful exfiltration of C-17 secrets between two PLA officers and another member of the hacking team—most likely a civilian employed as a PLA contractor. The report expressed the elation in having stolen secret information about a development project that had cost American taxpayers around $ 40 billion to develop between the 1980s to the 1990s.[111]

According to PLA's remarkably detailed summary of the operation revealed in federal court documents connected with the arrest and conviction of a major cyberespionage actor named Su Bin, aka Stephen SU.

Su was the owner of Beijing Lode Technology Company Ltd., an aviation and space technology supply firm with clients in China and around the world, including the United States.

The summary document boasted of great PLA satisfaction with its "Globemaster" operation:

> *In 2009, ... [we] began reconnaissance of C-17*
> *strategic transport aircraft, manufactured by the*
> *American Boeing Company and code-named*
> *"Globemaster." ... [W] e safely, smoothly*
> *accomplished the entrusted mission in one year,*

making important contributions to our national defense scientific research development and receiving unanimous favorable comments. . . .

The development of C-17 strategic transport aircraft is one of the most time-consuming projects in the American history of aviation research and manufacture, a total of 14 years from 1981 when the McDonnell Douglas Company won the development contract to 1995 when all test flights were completed. In development expenses, it is the third most expensive military aircraft in American history, costing the U.S. $ 3.4 billion in research and development.

Thorough planning, meticulous preparations, seizing opportunity ..., [we] initiated all human and material preparations for the reconnaissance at the beginning of 2009. After a few months' hard work and untiring efforts, through internal coordination [we] for the first time broke through the internal network of the Boeing Company in January of 2010.

Through investigation of Boeing Company's internal network, we discovered that the Boeing Company's internal network structure is extremely complex. Its border deployment has FW and IPS, the core network deployment has IDS, and the secret network has ... [a] type isolation equipment an anti-invasion security equipment in huge quantities. Currently, we have discovered in its internal network 18 domains and about 10,000 machines.

Our reconnaissance became extremely cautious because of the highly complex nature of Boeing's internal network. Through painstaking labor and slow groping, we finally discovered C-17 strategic transport aircraft-related materials stored in the secret network. Since the secret network is not open 24 hours and is normally physically isolated, it can be connected only when C-17 project-related personnel has verified their secret code.

Because we were well-prepared, we obtained in a short time that the server's file list and downloaded a small number of documents. Experts have confirmed that the documents were truly C-17 related and the data scope involved the landing gear, flight control system, and airdrop system, etc. Experts inside China have a high opinion about them, expressing that the C-17 data were the first ever seen in the country and confirming the documents' value and their unique nature in China.

Scientific/ technical support, safely procure clear achievement. Since the Boeing Company's internal network structure is highly complex and strictly guarded, successful procurement of C-17 related data required meticulous planning and vigorous technical support. We were able to deal with them one by one in our work.

[1] We raised the difficulty level of its counter reconnaissance work to ensure the secure obtainment of intelligence. From breaking into its internal network to obtaining intelligence, we repeatedly skipped around in its internal network to make it harder to detect reconnaissance, and we also skipped around at suitable times in countries outside the US. In the process of skipping, we were supported by a prodigious quantity of tools, routes, and the process of skipping, we were supported by

a prodigious quantity of tools, routes, and servers, which also ensured the smooth landing of intelligence data.

(2) We used technology to exit the network securely. Because breaking into Boeing's internal network was harder than we imagined, after obtaining intelligence we had to rely on technology to separate and bundle data, change the document formats, etc. Ultimately, we avoided the many internal automatic and manual auditing facilities to transfer data safely and smoothly out of the Boeing Company.

(3) We repeatedly skipped around to retreat safely. To ensure obtaining intelligence safely and evading tracking by American law enforcement, we had planned for numerous skip routes in many countries. The routes went through at least three countries, and we ensured one of them did not have friendly relations with the US. To safely, smoothly accomplish this mission, we opened five special routes and servers outside the U.S. and shut them down after the mission concluded.

(4) We made appropriate investments and reaped enormous achievements. Through our reconnaissance on the C-17 strategic transport aircraft, we obtained files amounting to 65G [gigabytes]. Of these, there were 630,000 files and 85,000 file folders, containing the scans of C-17 strategic transport aircraft drawings, revisions, and group signatures, etc. The drawings include the aircraft front, middle, and back; wings; horizontal

stabilizer; rudder; and engine pylon.

The contents include assembly drawings, parts and spare parts. Some of the drawings contain measurement and allowance, as well as details of different pipelines, electric cable wiring, and equipment installation.

Additionally, there were flight test documents. This set of documents contains detailed contents, and the file system is clear and detailed, considered topflight drawings by experts! This project took one year and 2.7 million RMB to execute, showing cost-effectiveness and enormous achievement. This reconnaissance job, because of the ... sufficient preparations, meticulous planning, has accrued rich experience for our work in the future.

We are confident and able ... to complete a new mission .... August 6, 2012.

As noted by Bill Gertz, less than a decade after the Boeing C-17 data heist, the Chinese showcased their nearly identical Xian Y-20 heavy transport jet at a November 2018 Zhuhai International Air Show. Their propaganda outlets bragged that the Y-20 "made China the third country after Russia and the U.S. to design and develop its own heavy military transport aircraft."

The first Y-20 prototypes were being built in 2013—just three years after the Boeing hack.[112]

Whereas Boeing's C-17 aircraft were built at an average cost of $202 million each, thanks to billions of dollars in design and manufacturing secrets stolen, the Y-20 was marketed by the state-run Aviation Corporation of China (known as AVIC) for only about $160 million.

Again, public federal court documents connected with the SU Case reflected the Chinese even more about the Chinese military's relentless drive to steal U.S. weapons knowhow.

An internal PLA document titled *F-35 Flight Test Plan* stolen from Lockheed Martin revealed secrets about America's most advanced stealth jet fighter bomber. The plan not only described how the F-35s would be tested as it was being developed, but also

> ...how many airplanes would be built and used; how certain components would be tested, how they could be configured, and using what instrumentation; and the techniques used to test the performance, capabilities, and limits of various features of the F-35.[113]

Court papers included the testimony of John Korstan, a former LF-35 principal designer at Lockheed Martin who had developed the plan, stated that:

> The F-35 was developed over multiple years by multiple companies performing contracts with the United States Department of Defense in the United States and other countries at a total cost of billions of dollars... Approximately 59,959 man-hours were required to create the F-35 Flight Test Plan.[114]

Korstan was shown Chinese flight plan translation produced by Su which included images of the document he had authored which included a "Cooperative Avionics Test Bed" (CATBird) used inside a Boeing 737 jet to test the F-35's avionics. Outside access to such sensitive technical F-35 information had been protected through the use of login credentials and by restricting physical access to facilities where the confidential information was stored.

As Korstan 's written testimony confirmed: "From my work at [Lockheed Martin] Aero, I know that F-35 flight test information is not publicly available." [115]

The Lockheed Martin hacking operation code-named by NSA as Byzantine Hades/Anchor was based in Chengdu Province. That information was then supplied to another AVIC subsidiary, the Chengdu Aircraft Industry Group, which used the stolen data in developing and testing a Chinese J-20 stealth fighter jet which featured F-35 radar-avoidance devices.

Unveiled in 2011, the stolen F-35 design features included a new electro-optical targeting pod deployed under the jet's nose cone, concealment of previously protruding engine nozzles at the tail to reduce the jet's radar signature, and a new radar-absorbing airframe coating.

A January 2014 Chinese Communist Party-affiliated Global Times newspaper even publicly boasted that key J-20 technologies had been obtained from the F-35.[116]

Washington Times national security columnist Bill Gertz's *Deceiving the Sky* reported that the Su case also revealed another damaging compromise of valuable technology related to the development of America's F-22 stealth aircraft. Included are details regarding its radar-invisible internal missile storage bay and a unique "supercruise" capability that permits fuel-efficient long-range sortie missions with enough fuel for safe return to base.

Court documents revealed that Su and 3PLA hackers had obtained copies of a presentation document developed by an aircraft company called EDO which had participated in the development of the F-22's missile launch bay known as the "Vertical Eject Launcher" ("AVEL"), along with an AMRAAM" (short for "advanced medium-range-air-to-air missile").

The presentation included three-dimensional renderings, photographs of the AVEL and its components, a cross-section diagram, mechanical design schematics, hydraulic/ pneumatic schematics, electrical schematics, and other details and descriptions of functions of AVEL parts.

The document also listed key quantitative metrics for the launcher's performance and measurements on the strength of components, as well as instructions for its installation and removal.

Nicholas DeSimini, a former EDO systems engineer, explained on record that AVEL applied a novel "trapeze" system of metal arms to eject missiles from the weapons bay that avoided a requirement for explosive release mechanisms used for radar-opaque underwing-mounted missiles. Although general drawings of the system had been made public in the early 2000s, they did not give of information provided in EDO's presentation.

DeSimini said, "Through an examination of the illustrations and written descriptions, a person would be able to reverse-engineer much of the work that went into creating the AVEL in the first place." [117]

The Chinese did exactly that. In the summer of 2013, a Chinese military enthusiast website disclosed a nearly identical vertical missile launcher incorporated into their J-20 fighter.

Su Bin and the 3PLA spies that had operated undetected for six years were eventually identified in 2013 through a secret NSA cyber counterintelligence program called "ArrowEclipse" that has tracked numerous Byzantine Hades operations and its offshoots. "Byzantine Candor", for example, involved hacks against Pentagon computers; "Byzantine Raptor" additionally targeted computers at the Pentagon and in Congress; and "Byzantine Foothold" hacked the U.S. Pacific Command and U.S. Transportation Command networks. [118]

Active PLA cyber espionage operations which began in the early 2000s, continued with no signs of abating. Bill Gertz reports that a previously classified made public in 2013 revealed that NSA had detected more than 30,000 incidents related to Byzantine Hades. At least five hundred of those events were described as significant intrusions of the Pentagon and other computer systems.

As a consequence, more than 1,600 network computers were penetrated, compromising 600,000 user accounts, and causing more than $ 100 million in damage to rebuild the networks. [119]

Michelle Van Cleave, a former high-ranking U.S. counterintelligence official within the Office of the Director of

National Intelligence, said while the Su (six-year jail sentence) prosecution was successful, the case only represented a drop in a bucket in stemming the torrent of Chinese thefts of American of secrets that keeps growing bigger every year.

Van Cleave said "What they can't acquire legally through trade, or creatively through mergers and acquisitions, they are prepared to steal. And it's getting harder all the time to stop them." [120]

Greatly compounding electronic theft problem, Chinese information and technology thieves don't have to depend on cyber espionage that involves hacking their way into computer networks. They can—and do—insert covert eavesdropping and control bugs into systems we and others purchase from them. [121]

On a summer evening in 2004, as the Supercomm tech conference in Chicago wound down, a security guard stopped a middle-aged fellow who was wending his way through unattended booths, popping open million-dollar networking equipment, and photographing internal circuit boards. [122]

The man, an engineer, identified himself as Zhu Yibin, a Chinese citizen wearing a conference lanyard that read "Weihua." He explained that this spelling was an accidental scramble of his employer's name: Huawei Technologies Co.

Security staff confiscated memory sticks with photos, a notebook with diagrams and data belonging to AT&T Inc., and a list of six companies including Fujitsu Network Communications Inc. and Nortel Networks Corp.

The engineer appeared for further questioning the following day. Rumpled and bewildered, he claimed that it was his first time in the U.S. and incredulously wasn't familiar with Superccomm's rules forbidding photography. [123]

Someone in Huawei's U.S. offices should have informed Zhu Yibin in advance that such prying practices are frowned upon.

The company opened offices in Plano, Texas (2001) and Santa Clara, Calif. (2002) which contained its offices were spy-proof secure rooms that were off-limits to American employees.

Counterintelligence officials have long suspected that Huawei (pronounced WAH-way) has been conducting state intelligence services through a secret and protected communications channel. U.S. firms doing business

While U.S. firms doing business with Chinese companies are concerned about tech theft, the potential riches have persuaded many to forgo making official complaints about commercial-secrets theft, even though many have privately sought help from U.S. officials.

In February 2007, U.S. customs officials arrested a Huawei employee and former software engineer for Motorola, Hanjuang Jin, at Chicago O'Hare International Airport. The bag she was carrying contained more than 1,000 proprietary Motorola documents, along with a one-way ticket to China.

The U.S. convicted Ms. Jin in 2012 on charges of stealing trade secrets.[124]

Meanwhile, Huawei has rapidly grown from a little-known interloper into China's global tech champion. The company employs 188,000 people in more than 170 countries, sells more smartphones than Apple—provides cloud services, makes microchips and runs under-sea cables that ferry global Internet traffic.

Huawei has also become the world's biggest maker of telecoms gear and a world leader in developing next-generation 5G networks. Once-mighty rivals such as Cisco and Motorola are now only a fraction of Huawei's size. According to S&P Global Market Intelligence data, only Google, Amazon.com Inc., and Samsung Electronics Co have larger research budgets.[125]

American intelligence experts regard Huawei to be a huge United States security threat; a company that implants secret "backdoor" malware into its diverse sales products ranging from cell phones to giant switches that run telephone networks to corporate computer systems.

Classified intelligence reports and unclassified congressional studies warn that the People's Liberation Army and China's Ministry of State will someday exploit those back doors to get inside American networks.[126]

A 2005 RAND Corporation investigation sponsored by the U.S. Air Force identified Huawei to be high on the list of Chinese networking firms considered to present security threats. The RAND study concluded that a "digital triangle" of Chinese firms, the military, and state-run research groups were working together to bore deeply into the networks that keep the United States and its allies running.

At the center of the action, the RAND report suggested, was Huawei's founder, Ren Zhengfei, a former PLA engineer who appeared to have never left his previous job.

The Pentagon has warned that China has already attempted to probe U.S. networks for data useful in any future crisis. The Department of Defense's 2017 annual report to Congress stated:

> *Targeted information could enable PLA cyber forces to build an operational picture of U.S. defense networks, military disposition, logistics, and related military capabilities that could be exploited before or during a crisis.*[127,128]

Short of banning U.S. companies from doing business with Huawei, the word has gone out that they do so at their peril. Although lacking conclusive evidence, the U.S. Congress concluded that Huawei and another Chinese company, ZTE, must be blocked from "acquisitions, takeover or merger" in the United States and "cannot be trusted to be free of foreign state influence."

Washington drew a firewall around the United States when Huawei attempted to buy 3Com, a failing American firm. The Committee on Investments in the United States—a little-known government agency run as an offshoot of the Treasury Department—blocked the purchase on national security grounds.[129]

US officials believe the Chinese government could use Huawei telecom gear to spy or disrupt communications. Many are particularly worried about future Huawei penetrations of government 5G networks.[130]

Chinese national security threat concerns aren't limited Huawei. After Lenovo, a Chinese computer upstart bought IBM's computer division in 2005, the State Department and the Pentagon largely banned its indestructible laptops.

The Office of Management and Budget (OMB) issued an interim rule in May 2019 laying out steps to ensure U.S. government agencies aren't doing business with Huawei and several other Chinese companies. The rule then allows agencies another year to comply with a separate, more restrictive provision that would bar the government from contracting with any company that uses products or services from Huawei or other banned companies.[131]

OMB's action implements policies of a 2018 National Defense Authorization Act (NDAA) that specifically targets Huawei and other Chinese tech companies including Huawei rival ZTE Corp. and surveillance equipment maker Hangzhou Hikvision Digital Technology Co. which are considered to present security risks. These provisions will ban U.S. agencies and recipients of federal grants and loans that make substantial use of their products from doing business with the Chinese companies or contractors.

Other Trump administration restrictions separate from the NDAA prevent trade-blacklisted companies, including Huawei, from receiving certain security-sensitive US-sourced products without special licenses. The U.S. Commerce Department is determining which U.S. company license applications for shipments to Huawei do and don't threaten national security.

Added to direct cybersecurity threats posed by Beijing and its information technology giants, China also continues to exert ever-expanding cyber assaults upon American and global commercial entities and major social media providers through the weaponization of the "Internet of Things." Much more discussion on this topic will follow later.

North Korea: Hermit Hackers and Hostilities

Just before the 2009 July Fourth holiday, a North Korean agent sent out a coded message with a simple set of instructions to about 40,000 computers around the world that were infected with a botnet virus. Each of the zombie computers was told to begin pinging a list of U.S. and South Korean government websites and international companies.

Almost immediately, the websites were flooded with more requests than their servers could handle and shut down. Some U.S. sites were hit with as many as one million requests per second.

Between July 4 and July 9, the U.S. Department of Homeland Security's dhs.gov along with state.gov websites became temporarily unavailable. So did those of the U.S. Treasury, Secret Service, Federal Trade Commission, Department of Transportation, and other agencies. The NASDAQ, New York Mercantile, and New York Stock Exchange sites were also hit, as was the *Washington Post*.[132]

However, the Distributed Denial of Service (DDoS) attack aimed at the White house failed. Only sites hosting the White House in Asia had trouble.

America was but one of the many cyber targets. On July 8, another 30,000 to 60,000 computers infected with a variant of the virus attacked a dozen or more South Korean government sites, banks, and an Internet security company.

The final July 10 BotNet assault which began at 6:00 p.m. Korea time enlisted an estimated 166,000 zombie computers in seventy-four counties in flooding more South Korean bank and government sites with unwelcome server requests.[133]

The 2009 DDoS messages were sent as a warning to the world regarding North Korean cyber capabilities. Although expansive, the attack didn't attempt to gain control of any government systems, disrupt any essential services, nor physically damage equipment systems.

And although the Obama administration chose not directly

attribute the attack to North Korea, the South Korean government was not similarly shy in doing so.

South Korea's National Intelligence Service (NIS) said in a statement that it had evidence that pointed to North Korea which suspiciously pointed to military preparations to destroy South Korea's communications infrastructure. NIS maintained that a North Korean hacker unit known as "Lab 110," or the "technology reconnaissance team," was ordered to prepare a plan the previous month to "destroy the South Korean puppet communications network in an instant."

A Vietnamese firm, Bach Khoa International Security (BKIS), traced the plan implementation back through eight servers controlled from a server in Brighton, England. From there, the trail went cold. This assessment was endorsed by the Korean Communications Commission.[134]

Unit 119 is only one of the hermit regime's four cyberwarfare organizations operated by the Korean People's Army (KPA). According to a former hacker who defected in 2004, Joint Chiefs Cyber Warfare Unit 121, by far the largest and best trained with over 600 hackers. This unit specializes in disabling South Korea's military command, control, and communications networks.

The North's Enemy Secret Department Cyber Psychological Warfare Unit 204 has 100 hackers and specializes in cyber elements of information warfare. Another Central Party's Investigations Department Unit 35 is a smaller but highly capable cyber unit with both internal security functions and external offensive cyber capabilities.[135]

Because Internet connections in North Korea are few and readily identified, many of their cyber operatives are stationed in China.

Four floors of the Shanghai Hotel in the Chinese town of Dandong on the North Korean border are reportedly rented out to Unit 110 agents. Another unit occupies several floors in the Myohyang Hotel located in the town of Sunyang.[136]

Selection and grooming for North Korea's elite hackers began early at the elementary-school level. They become trained in computer programming and hardware during middle and high school years. Upon graduation, they become automatically enrolled at the Command Automation University in Pyongyang, where their sole academic focus is to learn how to hack into enemy network systems.

The U.S. CYBER COMMAND ranks North Korea's capabilities along with China, Russia and Iran as top strategic threats to U.S. national security, energy grids, and financial stability.

In October 2009, only three months after DDoS attacks, South Korean media outlets reported that hackers, believed to be North Koreans, had infiltrated the Chemicals Accident Response Information System and withdrew a substantial amount of classified information on 1,350 hazardous materials.

It took seven months before the hack was discovered and attributed to malicious code covertly or accidentally implanted in the computer system by a South Korean army office. As a result, North Korea learned how and where South Korea stores its hazardous gases, including deadly chlorine used for water purification.

North Korean cyberattacks on energy grids and Internet communications—both in South Korea and the U.S.—are obvious threats. North Korea has far less of either to worry about concerning retaliation in kind.

North Korea barely has an electric grid, and fewer than 20,000 of North Korea's 23 million citizens even have cell phones. Radios and TVs are hardwired to tune only into official government channels. And as for the Internet, North Korea is characterized as a "black hole."

As David Sanger notes in his book, *The Perfect Weapon,* North Korea has discovered the true merits of weaponizing the Internet both to wreak havoc and to make profits.

Kim Heung-Kwan, a North Korean defector who had helped train many of the North's cyber spies, told the *New York Times* about a "very strange idea" that a groups of returning group of North Korean computer experts brought back from assignments in China in the early 1990s. That strange idea was to use the Internet to steal secrets and attack the government's enemies.

"The Chinese are already doing it," he remembered one of the experts said.[137]

The North Korean military reportedly began training computer "warriors" in earnest in 1996. Bureau 121 was the first unit to open for business following two years of training in China and Russia.

Jang Sae-Yul, a former North Korean army programmer who defected in 2007, said these prototypical hackers were envied, in part because of their freedom to travel.

"They used to come back with exotic foreign clothes and expensive electronics like rice cookers and cameras," he said.[138]

Jang, a former first lieutenant in an army unit that wrote software for war game simulations, reported that his friends told him that Bureau 121 was divided into different groups, each targeting a specific country or region. One group, for example, focused on the United States, and another on South Korea. There was even one focusing on the North's lone ally, China.

As Jang reported: "They spend those two years not attacking, but just learning about their target country's Internet."

As time went on, Jang said, the North began diverting high school students with the best math skills into a handful of top universities. The one he attended as a young army officer was a military school called Mirim University that specializes in computer-based warfare.

Other students were deployed to an "attack base" in the northeastern Chinese city of Shenyang, where there are many North Korean-run hotels and restaurants.[139]

The U.S. government intelligence officials grossly underestimated how rapidly North Korea would advance their military hacking prowess, just as they have done regarding the "backward" nations forward progress in long-range missile developments. A 2009 NSF Intelligence Estimate had concluded that it would take years before North Korea could mount any significant threat.

That U.S. complacency soon began to change. The FBI's counterintelligence division, for example, started to notice that many North Koreans assigned to work at the United Nations were also quietly enrolling in university computer programming courses in New York.[140]

North Korean hackers also likely learned skills from Iran. Both countries have long-shared both missile technology and a common view of the United States is the major source of their problems.

The North also learned something else of great importance from Iran: when confronting an enemy that has Internet-connected banks, trading systems, oil and water pipelines, dams, hospitals, and entire cities, the opportunities to cause troubles are endless.

In 2013, Pyongyang launched its first big cyber strikes conducted against computer networks at three major South Korean banks and two of South Korea's largest broadcasters... just seven months after Iran, struck Saudi Aramco using a very similar cyber weapon. The WannaCry hack occurred during joint American and South Korean military exercises.

As with Iran's Saudi attacks, the North Korean operation on South Korean targets—dubbed by the U.S. as "Dark Seoul"—WannaCry used wiping malware to eradicate and paralyze business operations.

When a young 27-year-old Kim Jong-un succeeded his father Kim Jong-il as North Korean leader in 2011, the rapidity in which he asserted full control over the military surprised everyone. As David Sanger observes, Kim's priority was to make North Korea's nuclear threat a credible one. His second was to eliminate potential rivals, which he sometimes did with an anti-aircraft gun. His third was to build a cyber force, and he brought to this task a sense of urgency.

Kim followed his father's footsteps in recognizing the strategic importance of cyber espionage and warfare. Kim Jong-il reportedly told top commanders in 2003...

> *If warfare was about bullets and oil until now,*
> *warfare in the twenty-first century is about*
> *information.* [141]

Kin Jong-un wasted no time targeting cyberattacks on foreign military intelligence sources, banking systems, and currency exchanges.

As tallied in a September 16, the *Wall Street Journal* report, a partial chronology of North Korean operations includes:

- December 2016: Emails stolen from Sony Pictures Entertainment.

- February 2016: $81 million stolen from Bangladesh central bank.

- September 2016: South Korean defense minister's personal computer hacked for military intelligence.

- May 2017: WannaCry ransomware attack infects more than 300,000 computers in 150 countries.

- November 2017: Adobe Flash "zero-day" malware is embedded in Microsoft Office files in South Korea.

- December 2017: Attacks on South Korean groups affiliated with the Winter Olympics.

- December 2017: South Korea cryptocurrency exchange Youbit is hacked, causing it to declare bankruptcy.

- January 2018: Tokyo-based Coincheck cryptocurrency exchange reports about $530 million was stolen.

- March 2018: Adobe Flash zero-day attack on Turkish

financial institutions and government groups.

- March 2019: $49 million stolen from a Kuwait institution.

The December 2016 attack on Sony Pictures Entertainment in retaliation an upcoming and very unflattering farcical Hollywood movie titled *"The Interview"* for reflected the importance Kim attached to the portrayal of his regime its leadership in the global media.

It isn't difficult to understand why Kim didn't appreciate the satirical humor.

The movie plot featured two bumbling, incompetent journalists played by actors Seth Rogen and James Franco who had arranged to interview Kim Jong-un being recruited by the CIA to assassinate the dictator by blowing him up.

Forewarned about the plot, North Korea's foreign ministry sent a letter of protest to UN Secretary-General Ban-Ki-moon seeking his intervention to block the movie's distribution.

It likely didn't work in favor of the request for intervention that the secretary general, a South Korean, wasn't particularly inclined to solve the North's problem. In any case, neither Ban-Ki-moon, nor the UN was in a position to influence a commercial Hollywood studio decision.

Due to receiving little or no response, Kim's representatives began issuing direct threats against the United States. The missives warned that Sony's planned Christmas day 2014 release of the movie would be viewed as an "act of terrorism" meriting "a decisive and merciless countermeasure."

Sony Corporation's Tokyo-based top executives took the threats very seriously, ordering the U.S. Sony Pictures Entertainment studio to, at least, tone down a scene at the end of the movie in which Kim's head appears to explode during a gruesome assignation. Wishing to further distance themselves from the contentious controversy, Sony's corporate leadership in Tokyo removed the name "Sony Pictures" from all of the film's posters and promotional materials.

It took months—until December 2017, three years to the day after Obama accused North Korea of the Sony attacks—for the United States and Britain to formally declare that Kim Jong-un's government was responsible for WannaCry.

By the time *"The Interview"* was being made, the Hermit Kingdom had gone from viewing the Internet as a threat to viewing it as a marvelous invention for enormously lucrative cyber theft and intimidation.

The U.S. and UN experts believe that cyber-enabled heists have become a crucial North Korean revenue stream and security threat that could soon rival its weapons program.

In February 2016, North Korean hackers got inside the website of Poland's financial regulator and infected visitors. They also attempted to break into central banks of Venezuela, Estonia, Brazil, and Mexico.

Two of Pyongyang's boldest attacks occurred in fall of that year—one on South Korea, the other on the world. The cyber assault breached South Korea's Defense Integrated Data Center. According to Rhee Cheol-hee, a member of the South Korean parliament's National Defense Committee, the penetration swept 182 gigabytes of data—including OpPlan 5015, a detailed outline of what the U.S. military delicately called a "decapitation strike." [142]

OpPlan 5015 appears to have included strategies for finding and killing the country's top civilian and military leaders and seizing as many nuclear weapons as possible...and it didn't stop there. The strategy also included ways to counter the North's elite commandos, who would almost certainly slip into the South.

South Korea's Defense Integrated Data Center penetration demonstrates how deeply the North has compromised South Korea's sensitive networks. There is also evidence that Pyongyang has planted "digital sleeper cells" in critical infrastructure in the South in case they are needed to paralyze power supplies or command-and-control systems. [143]

On May 12th 2017, North Korea WannaCry ransomware attack which hit the British Health Service and shut down businesses all over Europe and North America and demanding payments in Bitcoin to unlock their computer systems was undertaken by a hacking organization known to Western analysts as the Lazarus Group.

In September 2019, Treasury identified and blacklisted three hacking groups allegedly run by North Korea's primary intelligence service that were connected expansive operations across 10 countries. One such collective, the Lazarus Group along with its two subsidiaries, known as Bluenoroff and Andariel, reportedly stole $700 million in three years—and may have attempted to steal nearly $2 billion.

The Trump administration had previously blamed the Lazarus Group for the WannaCry worm, infecting more than 300.000 computers, crippling banks, hospitals, and other companies in 2017. The Justice Department later also charged North Korean operative Park Jin Hyok and unnamed co-conspirators with the 2016 WannaCry hack on Sony Pictures, and $81 million theft from Bangladesh's account in the Federal Reserve Bank of New York.

A single typo in the Bangladesh heist prevented the hackers from stealing the $851 million.

John Hultquist, director of intelligence analysis at the cybersecurity company FireEye Inc. is convinced that most of these cyber theft activities trace back to Pyongyang government fingerprints:

> *Though these operations may fund the hackers themselves, their sheer scale suggests that they are a financial lifeline for a regime that has long depended on illicit activities to fund itself.*[144]

Many cyber experts also believe that these highly lucrative high-value hacking operations help to keep Kim Jung-un's cash-hungry regime in power by insulating the North Korean economy from otherwise crushing global sanctions which are intended to force Pyongyang into giving up its weapons of mass destruction.

Investigations reveal that some of the money from that cyber theft gets channeled into the nuclear weapons and ballistic programs those U.S. and UN sanctions are intended to stop. This being the case, many cybersecurity experts believe it unlikely that Pyongyang will be pressured through sanctions into curtailing its malicious behavior.

Some industry experts hypothesize an opposite scenario; namely that Mr. Kim's increased willingness to talk about denuclearization over the past few years may arise from a belief that his country's cyber arsenal can partially supplant its weapons as a leveraging threat to other nations.

As North Korean experts Mathew Ha and David Maxwell observe in a Foundation for the Defense of Democracies report:

> *North Korea's cyber operations broaden the Kim family regime's toolkit for threatening the military, economic, and even the political strength of its adversaries and enemies.*[145]

At a time when the U.S. was busily planning ways about how to use cyberweapons to neutralize the North's missiles, the North was thinking about how to use them to pay for those missiles—a huge challenge for a country under every form of economic sanction.

Two years after their Sony attack, they demonstrated a scheme to do just that—one planned to extract $1 billion in 2016 from the Bangladesh Central bank.[146]

Following a few weeks of quiet electronic bank observation, the hackers got all the bank digital lockbox withdrawal needed: procedures for transferring funds internationally, some stolen credentials, and an understanding of when the bank would be closed for a holiday that extended into a weekend. The extra days provided them with time to execute transfers before anyone was around to stop them.

The hackers' illicit bank transfer orders amounted to just under $1 billion. One, a transfer to the Shalika Foundation in Sri Lanka, incorporated a small spelling error that led to a big mistake.

In instructions to the New York Federal Reserve, through which such transactions flow, someone spelled "foundation" as "fandation." The error raised eyebrows, and the transfers were suspended—but only after Kim Jong-un's hackers had gotten away with $81 million.[147]

North Korean hacks of financial systems and critical infrastructure worldwide reveal increasingly sophisticated cyber capabilities designed to circumvent far easier to trace violations of global trade sanctions.

Nevertheless, USCYBERCOM, working together with the Department of Treasury and the Department of Homeland Security and Infrastructure Security Agency, has uncovered a variety of hidden worms with known Pyongyang addresses.

The UN has recently investigated at least 35 reported North Korean cyberattacks across five continents targeting banks, cryptocurrency exchanges, and mining companies. Their hackers are known to have attempted five major world-wide cyber thefts in 2019 alone, including a successful $49 million heist from a Kuwait institution.

In 2019, UN investigators who tallied proceeds from all reported Lazarus and other North Korean hacking groups estimated thefts in billions of dollars.

US security officials and cyber experts believe that most estimates likely underestimate real losses because many known thefts bearing hallmarks of North Korean involvement aren't publicly reported due to fear of embarrassment and further exposure.[148]

Under Trump administration policies, U.S. security agencies are mandated to immediately disclose malware samples they discover with private industry to enable timely countermeasure protections. In September 2019, for example, a North Korea

"Hidden Cobra," alert was dispatched regarding a new malware discovery dubbed "ELECTRICFISH" that burrows into computers to steal data.[149]

Iran: Shamoon Invades Aramco

Although Iran is believed, by many cybersecurity experts to possess less immediate ability to penetrate critical U.S. banking and other infrastructure networks than Russia, China or North Korea, there are clear signs that they are trying to do so.

As reported by Annie Fixler, deputy director of the Center on Cyber and Technology Innovation at the Foundation for the Defense of Democracies, it is a mistake to dismiss Tehran's ability to conduct significant cyber operations.[150]

Whereas the January 2019 update of the annual Worldwide Threat Assessment report published by the U.S. intelligence community concluded that Iranian hackers are only capable of "causing localized, temporary disruptive effects," the cybersecurity firm FireEye warned that same month that Iranian operations pose a threat to "a wide variety of sectors on a global scale."

A European Union report also released in January 2019 warned that Iran will likely "intensify state-sponsored cyber threat activities." That observation appears to fit and observed an accelerating pattern.[151]

Between 2011 and 2013, Iranian hackers knocked Saudi Aramco's business offline for months and cost U.S. financial institutions tens of millions of dollars. Between 20017 and 2019, they hit more than 200 companies around the world and inflicted hundreds of millions worth of damage.

The August 2012 cyber attack during Ramadan against Saudi Aramco, which Tehran viewed as America's "gas station," found an easy target. Knowing that most of the company workers would be away, hackers flipped a switch that unleashed a simple wiper virus called "Shamoon" onto 30,000 Aramco computers and 10,000 servers. Screens went black, and files disappeared. Some screens reappeared showing a partial image of a burning American flag.[152]

Panicking Saudi technicians responded by ripping cables out of their computer servers, physically unplugging Aramco offices around the world. And although oil production wasn't affected, most everything surrounding it was severely disrupted.

Supply purchase data was wiped out, as were means to coordinate product shipping operations. Officials and workers lacked contact with the Saudi Ministry of Energy, with oil rigs, or with the giant Kharg Island oil terminal, through which the Saudis ship much of their crude production. There was no corporate email and the phones were dead.[153]

The attack—was soon traced in part to a Saudi Aramco insider—forced the Saudis to scrap and replace their infected computers. It took five months to undo the costly damage.

That, however, was only the beginning.

A 2017 Iranian hacking attack against the British parliament compromised dozens of email accounts belonging to lawmakers by identifying weak passwords lacking two-factor authentication.[154]

Researchers at US-based firms have reportedly observed numerous Iranian state-sponsored hacking attempts—particularly through spear-phishing attempts—against U.S. government and private industry targets. These attempts appear to be tied to escalating tensions between the U.S. and Iran, although not successful.[155]

Investigators have noted that much of the targeting appeared to be focused on U.S. government and energy sector entities, including oil and gas. Some email lures were posted as messages from the White House's Executive Office of the President.

FireEye and two other US-based cybersecurity firms, CrowdStrike Inc. and Dragos Inc., said the Iranian phishing campaign appeared linked to a known Iranian hacking group believed to possess powerful, destructive tools.[156]

In February 2019, the U.S. Justice Department unsealed an indictment against an American citizen and four Iranian operatives caught targeting U.S. government and intelligence assets through phishing attempts using fake Facebook profiles to trick victims into accepting friend requests. Fortunately, the phishing emails were poorly written with spelling errors that tripped up the tricksters.[157]

The Israeli military's former cyber chief Brigadier General Noam Sha'ar reported a failed Iran cyber attempt to infiltrate Israel's home front missile warning system. Had the penetration been successful, the hackers could have activated false alerts and/or preventing the system from detecting incoming rockets to warn citizens to take cover.[158]

In July 2019, suspected Iranian hackers infiltrated critical infrastructure and government computers in the Persian Gulf nation of Bahrain, a small but strategically important country. Bahrain is the permanent home of the U.S. Navy's Fifth Fleet and Central Command and is closely allied with its much larger neighbor, Saudi Arabia, a regional rival of Iran.

Unlike most Gulf states, where Sunni Islam is the dominant branch, Bahrain's population is about 70% Shia—the predominant faith in Iran—and its Sunni-led government for years has accused Iran of meddling in its affairs.[159]

By August, the hackers broke into systems of Bahrain's National Security Agency—the country's main criminal investigation authority—as well as the Ministry of Interior and the first deputy prime minister's office.

Bahrain authorities also identified intrusions into its Electricity and Water Authority were hackers who took over command and control shut down several systems. The operation is suspected to be a test run of Iran's capability to disrupt the country.[160]

The Bahrain breaches bear similarities to the devastating 2012 attack upon Saudi Aramco which unleashed the powerful Shamoon malware. Another attack traced to the same virus knocked Qatar's natural gas firm RasGas offline two weeks later.

RasGas is a joint operation of Qatar Petroleum and ExxonMobil is the world's largest producer of liquefied natural gas.

The company distributes about 36 million tons of resources annually.

Although traced most directly to the Russians, a 2017 penetration of the Triconex safety-instrumented system of a petrochemical plant in Saudi Arabia is suspected to have Iranian influences as well. Enormously sophisticated malware known both as "Triton" and "Trisis" used in the attack targeted industrial control systems and manipulated safety systems that could have caused lethal explosions.

Cyber investigators have traced a Triton hacker trail to probes on multiple North American and European oil and gas industry targets. Investigators believe that the tools and expertise required to develop the malware likely originated in a Moscow-based and government-owned and military-connected technical institute known as the Central Scientific Research Institute of Chemistry and Mechanics.

Based upon close geopolitical relationships with Russian, Iran is broadly suspected as complicit.

Regional Gulf leaders and U.S. security officials believe that Iran has been ratcheting up its malicious cyber activities in response to tensions over sanctions aimed at blocking its military nuclear ambitions.

According to a representative of Bahrain's Ministry of Interior:

> *In the first half of 2019, the Information & Government Authority successfully intercepted over 6 million attacks and over 830,000 malicious emails. The attempted attacks did not result in downtime or disruption of government services.*[161]

With or without Russia, concerns that Tehran is laying the groundwork for substantial infrastructure attacks against America and our allies are prudently warranted.

For example, the U.S. indicted seven Iranian nationals in 2016 that were allegedly working on behalf of the Islamic Revolutionary

Guard Corps to carry out intrusions that included the breach of a small New York dam.[162]

Tehran first publicly announced the creation of a Cybercorps in the summer of 2010, soon after the US-Israeli Olympic Games Stuxnet attack on Iran's nuclear program was exposed.[163]

As reported in the *Wall Street Journal*, Iranian hackers who appeared to be a network of fewer than 100 computer specialists at Iranian universities and network security companies soon hit a variety of commercial targets in an apparent payback for the Olympic Games.

To recapitulate:

- January 2012: Potent but relatively small-scale DDoS attacks against U.S. banks.

- July 2012: The Shamoon attack on Saudi Aramco's website and email system.

- August 2012: The attack against the Qatari RasGas website and email system.

- September 2012: A group called "Qassam Cyber Fighters" launches DDoS strikes against Bank of America Corp, J.P. Morgan Chase & Co., U.S. Bancorp, PNC Financial Services Corp., Wells Fargo & Co., and the New York Stock Exchange.[164]

Richard Clarke and Robert Knake linked the cyberattacks to Iran's military through its Revolutionary Guard (IRGC) and its Ministry of Intelligence.[165]

David Sanger notes in his book *The Perfect Weapon* that the Obama White House felt a need to hide evidence that Iranians were behind the attacks and immediately classified such attribution. Congressional staff members were shuttled into secure conference rooms before being told that Iran was a certain culprit and was cautioned not to reveal this information to the public.[166]

The secret didn't last long, Banks urgently wanted to know

who had hacked them, and enlisted their private security teams to identify the culprits.

Still, most of the targeted financial institutions decided it was better to shut up than to admit the existence of the attacks out of concern for alarming their customers. The banking industry did, however, begin to invest billions of dollars in cyber protection expertise and technologies to prevent future losses.

HACKERS RELEASE A VIRAL EPIDEMIC

MARKETING OF CYBERSPACE espionage and malware secrets became virtually a global industry with the advent of a new website launched in 2006 which purported the goal of "exposing corruption and abuse around the world."

In April 2010, the WikiLeaks site published a provocatively titled video, *Collateral Murder,* depicting an edited, annotated video from a U.S. Army Apache attack helicopter firing on civilians, including two *Reuters* reporters, in Iraq.

The video released by Australian editor, publisher and open information activist Julian Assange was first displayed at a news conference at the National Press Club in Washington, DC.

WikiLeaks followed up the video in July and October 2010 by releasing immense troves of classified documents related to the wars in Afghanistan and Iraq. For the classified documents, Assange worked with the *New York Times,* the *Guardian,* and *Der Spiegel* to verify, analyze, and present the documents to the public.

Again, just a few months later, WikiLeaks dropped another

virtual bomb later known as "Cablegate." The avalanche of documents included 251,287 confidential U.S. State Department cables written by 271 American embassies and consulates in 180 countries from December 1966 to February 2010. While most addressed rather routine and boring stuff, there were also several embarrassing items such as what American ambassadors thought about their foreign counterparts.

Some communications were more than gossipy. One revealing, for example, that the United States had secretly eavesdropped on the UN Secretary-General in the lead up to the Iraq war.

Unsurprisingly, American officials condemned the release of these documents in strong language and began to hunt down the source of the leaks. The U.S. government even went so far as to order federal employees and contractors not to read the secret State Department documents posted online. The New York Times somewhat humorously described this folly as "a classic case of shutting the barn door after the horse has left." [167]

Chelsea Manning: A Bradass Miss-fit

An early 2010 exchange on AOL Instant Messenger launched one of the biggest incidents in cyber history:

> *bradass87: hypothetical question: if you had free reign [sic] over classified networks for long periods of time...say, 8-9 months...and you saw incredible things, awful things...things that belonged in the public domain, and not on some server stored in a dark room in Washington DC...what would you do?*

> *(12:21:24PM) bradass87: say...a database of half a million events during the Iraq war...from 2004 to 2009...with reports, date time groups, lat-lon*

locations, casualty figures...? Or 260,000 state department cables from embassies and consulates all over the world, explaining how the first world exploits the third, in detail, from an internal perspective?...

(12:26:09PM) bradass87: let's just say *someone* I know intimately well, has been penetrating U.S. classified networks, mining data like the ones described...and has been transferring that data from the classified networks over the "air gap" onto a commercial network computer...sorting the data, compressing it, encrypting it, and uploading it to a crazy white-haired Aussie who can't seem to stay in one country very long=L...

(1:31:43PM) bradass87: crazy white haired dude=Julian Assange

(12:33:05PM) bradass87: in other words...I've made a huge mess.168

Bradass87 was the online handle of U.S. army private first class Bradley Manning born in 1987. As he described himself in instant messages sent to another hacker turned journalist, "I'm an army intelligence analyst, deployed in eastern Baghdad, pending discharge for 'adjustment disorder' in lieu of 'gender identity disorder.'"

Manning later changed "his" name to Chelsea based upon "her" preferred gender identity.

Later investigations found that Manning fit poorly with other soldiers and had previously been reprimanded for disclosing too much information in video messages to her friends and family he posted on YouTube. In fact, he almost wasn't deployed to Iraq because his superiors had described him as "a risk to himself [or herself] and possibly others."

Nevertheless, an urgent need for intelligence workers prevailed, and Manning was sent to the war zone where he/she was even allowed to handle classified information.[169]

Manning's job was "to make sure that other intelligence analysts in his group had access to everything that they were entitled to see." That position gave him access to a huge range of data streams from across the government's computer networks.

An increasingly distraught Manning took it upon himself to decide that "information has to be free." He also took it upon himself to release it to WikiLeaks.

Manning took advantage of that opportunity in exploiting a Department of Defense failure to adequately "air gap" their networks from Internet connections. Although DoD had banned the use of USB storage devices, they had not closed off access to writable CD drives.

Manning later reported bringing CDs to work with music on them, then overwriting the music with file upon file of classified data. He/she wrote in his/her chat log:

> *I listened and lip-synced to Lady Gaga's Telephone while exfiltratrating* [sic] *possibly the largest data spillage in American history.*[170]

The leaking and releases Manning's documents on WikiLeaks was roundly condemned for putting U.S. security assets at risk. Chinese nationalist groups, for example, conducted an "online witch hunt," threatening violence against any China dissident linked in the cables to meetings with the U.S. embassy.

Despite this reactionary pressure, the WikiLeaks organization survived. The leaked documents are still available around the Web on dozens of mirror websites to anyone who wants to see them.

Moreover, after being convicted by a court-martial in 2013 for violations of the Espionage Act and other offenses, her sentence was commuted by President Obama to nearly seven years served in confinement before his/her arrest in May 2010.

Manning unsuccessfully challenged incumbent Democrat Maryland Ben Cardin for his U.S. Senate seat in 2018. She was held in contempt by a U.S. District Court judge and again jailed on March 8th 2019, for refusing to testify against Julian Assange.

To this day, the NSA has never had to account for the fact that it ignored so many warnings about its well-documented vulnerabilities to a new era of insider threats. Nor have they been called to task for leaving him so loosely supervised that he could download highly classified documents that had nothing to do with his assigned work simply as a system administrator.

Nor was this the last time such a momentous security breach would make WikiLeaks more notorious.

Edward Snowden Drops a Bombshell

The same year Manning was convicted, a huge new bombshell breach of highly classified NSA data was disclosed to the world. In June 2013, *The Guardian, The Washington Post, The New York Times,* and *Der Spiegel* posted articles provided to them by Edward Joseph Snowden, a former CIA employee and subcontractor for the firm Booz Allen Hamilton.

Snowden's trove of leaks revealed far more than occasional cyber snooping on friends and foes. Rather, it became immediately public to the entire world that the U.S. had tasked thousands of engineers and contractors, working under tight security, to build a range of experimental weapons.

Some of these weapons merely pierced foreign networks and offered another window into the deliberations and secret deals of adversaries and allies—basically a cyber-assisted form of traditional espionage.

The leaked media postings revealed numerous global surveillance programs, many of which were run by the NSA and the "Five Eyes Intelligence Alliance" with the cooperation of telecommunication companies and European governments.

The secret revelations that got some of the biggest headlines in the United States revolved around a single copy of a "Verizon order" from the Foreign Intelligence Surveillance Court. The single document revealed that the secret court had developed a legal theory that the USA PATRIOT Act—passed in the days after 9/11—could be interpreted to require Verizon and other IT carriers, like AT&T, to turn over the "metadata" for every call made into and out of the United States. And then, it added, for good measure, all calls "wholly within the United States." [171]

American citizens were shocked and outraged to learn that the United States government was collecting what a presidential commission later called "mass, undigested, nonpublic personal information" just in case it wanted to mine that data sometime in the future "for foreign intelligence purposes."

One program, code-named "PRISM," allowed court-approved, if limited, access to the online Google and Yahoo! Accounts of tens of millions of Americans. Another program, named "XKeyscore," offered sophisticated new methods for NSA to tap into and filter vast flows of global Internet data.[172]

Snowden's leaks revealed that the NSA was willing to break the encryption of cell phone data, and how it was undermining even the "virtual private networks"—or VPNs—that companies and many computer-savvy users had turned to in hopes of protecting their data.

Public and private debates raged over whether Snowden was a hero or a villain in revealing that the NSA had overstepped its bounds in sucking up vast amounts of data on Americans, virtually washing away any distinction between "domestic" and "foreign" communications.

Whether American patriot or pariah, Edward Snowden's personal life and professional background lends potential insights regarding internal information security vulnerabilities.

That life experience began in a North Carolina family steeped in military tradition. Snowden had even enlisted in the U.S. Army Special Forces training program before washing out.

In 2006, after bouncing around several colleges, Snowden filled a CIA position needed to fulfill their growing counterterrorism mission. His first assignment was an undercover telecommunications job in Geneva. Concluding that his special expertise would be more profitably rewarded in the private sector, he quit three years later.

After taking a position with Dell advising NSA on updating its computers, working inside the agency became his priority. Whatever his motive, he went to the risk and trouble of breaking into a government computer system to swipe the NSA's admission test. Then, armed with the answers, he aced the exam and received an NSA offer for a mid-level bureaucratic position with a disappointing mid-level salary to match.

Snowden then applied for and secured the next best thing; a position with Booz Allen Hamilton, the company that had participated in the design of many of NSA's most important computer systems and also provided the staff to keep them running.

Whether motivated by a sincere interest in supporting NSA programs, or as a story he put out to cover his tracks, Snowden then professed to hate hackers.

Responding online to a 2009 *New York Times* article reporting that President Bush had turned down an Israeli request for bunker-buster bombs to deal with Iran's nuclear program, he wrote: "HOLY SHIT, WTF NYTIMES, Are they TRYING to start a war? Jesus Christ, They're like WikiLeaks." [173]

Snowden added: "Who the fuck are the anonymous sources telling them this? Those people should be shot in the balls."[174]

Snowden was assigned to work in support of a NSA division within a complex in Hawaii, which is not far from Pearl Harbor and the U.S. Pacific Command. That was where NSA was deploying many of its very best cyber weapons against its most sensitive targets, including North Korea's intelligence services and China's People's Liberation Army.

104

The weapons ranged from new intelligence techniques that could leap air gaps to penetrate computers not connected to the Internet, along with computer implants that could be detonated in time of war to missiles and blind satellites.[175]

Having soon purportedly becoming disillusioned with the information he was privy to, Snowden had left his job at the NSA facility in Hawaii during the month before his document dumps and flew to Hong Kong. Exactly how many documents, PowerPoint slides, and databases Snowden copies and smuggled with him when he fled to Hong Kong is still a matter of dispute.

Although most were never published, we can imagine that they provided all of America's adversaries with valuable gifts of information about NSA's global operations and general methods. Fortunately for NSA, Snowden only accessed documents that describe the agency's programs, but not the specific sources or details of the tools that enable them.

In Europe, the Germans were not pleased to learn that NSA had operated a surveillance office on the rooftop of the American embassy overlooking the famed Brandenburg Gate.

Documents that appear to have come from a leaker other than Snowden suggested that Chancellor Angela Merkel's personal cell phone became tapped as an NSA target after she became party leader of the Christian Democratic Union half a decade earlier. It wasn't a cyber operation; but just plain old phone tapping.

In any case, Merkel was outraged. She told President Obama: "Spying among friends—that simply isn't done." [176]

On the contrary, it is done all the time, including Merkel's own intelligence agency, the BND. But Merkel was hardly the only target—the United States was also listening to the leaders of Mexico and Brazil.[177]

As for Edward Snowden, he was later charged by the U.S. Department of Justice with two counts of violating the Espionage Act of 1917 and theft of government property.

Despite having his passport cancelled, Snowden still managed to fly to Moscow's Sheremetyevo Airport where he was restricted to the airport terminal for over a month before moving to Moscow where he has been granted repeated visa residence extensions.[178]

The Penetration of the CIA's Vault 7

Although the U.S. technology industry secured a commitment to disclose security vulnerabilities from the Obama administration in the wake of Edward Snowden's NSA leaks, CIA disclosures turned out to be quite a different matter.

Whereas the U.S. government agreed to a "Vulnerabilities Equities Process" disclosure agreement in response to lobbying pressure by technology companies who were at risk of losing global customer markets over real or perceived hidden risks, WikiLeaks has since released a huge new trove of documents demonstrating otherwise.

Besides, as reported by WikiLeaks, Snowden's limited access to classified documents had contained only hints of far more expansive U.S. government cyber espionage and information disruption programs. Taken together along with the release of a new series of CIA leaks which began in March 2017, they exposed a vast scope and scale of NSA, CYBERCOM, and CIA activities.

The major theft and public exposure of highly sensitive and secret CIA documents from what WikiLeaks called "Vault 7" was particularly concerning.

These materials obtained through 2016 revealed numerous zero-day exploits of widely used software, including the products of Apple, Microsoft, and Samsung (e.g., allegedly a tool to listen to rooms in which Samsung televisions were installed, even when the televisions appeared to be turned off). Yet when the documents became public, Microsoft president Brad Smith complained that no one in the U.S. government had told them about the vulnerabilities.[179]

Many of the vulnerabilities exploited in the CIA's cyber arsenal are pervasive and are likely already found by rival intelligence

agencies and cyber-criminals.
As noted by WikiLeaks:

> *The same vulnerabilities exist for the population at-large, including the U.S. Cabinet, Congress, top CEOs, system administrators, security officers, and engineers. By hiding these security flaws from manufacturers like Apple and Google, the CIA ensures that it can hack everyone...*[180]

The first part of the series WikiLeaks called its "Year Zero" program comprised 8,761 documents and files taken through 2016 from an isolated, high-security network situated inside the CIA's Center for Cyber Intelligence in Langley, Virginia. According to WikiLeaks, the number of Year Zero Vault 7 pages obtained had already eclipsed the number published over the first three years of the Edward Snowden leaks.

As WikiLeaks reported on its website:

> *Recently, the CIA lost control of the majority of its hacking arsenal including malware, viruses, Trojans, weaponized zero-day exploits, malware remote control systems and associated documentation.*
>
> *This extraordinary collection, which amounts to more than several hundred million lines of code, gives its possessor the entire hacking capacity of the CIA. The archive appears to have been circulated among former U.S. government hackers and contractors in an unauthorized manner, one of whom has provided WikiLeaks with portions of the archive.*[181]

According to WikiLeaks, the CIA's global covert hacking program "weaponized exploits against a range of U.S. and European company products, including Apple's iPhone, Google's Android and Microsoft's Windows, and even Samsung TVs, which are turned into covert microphones."

WikiLeaks reported that the CIA's malware and hacking tools are being built by its Engineering Development Group (EDG). EDG is responsible for the development, testing and operational support of CIA malware used in covert operations world-wide.

The CIA's Embedded Devices Branch (EDB) developed a "Weeping Angel" bug that infest smart TVs (Samsung in particular), transforming them into covert microphones. It accomplishes this by placing the target TV in a "Fake-Off" mode so that the owner falsely believes it is off when it is on. In "Fake-Off" mode, the TV records room conversations which it transmits over the Internet to a covert CIA server.

The CIA's Mobile Devices Branch (MDB) has developed numerous attacks to remotely hack and control popular smartphones, instructing them to send the CIA the user's geolocation, audio and text communications as well as covertly activate the phone's camera and microphone.

WikiLeaks notes that despite the iPhone's minority share (14.5%) of the global smartphone market in 2016, a specialized unit in the CIA's Mobile Development Branch has produced malware to infest, control and infiltrate data from iPhones and other Apple products running iOS, such as iPads. They reported, "The disproportionate focus on iOS may be explained by the popularity of iPhone among social, political, diplomatic and business elites."

A similar CIA unit targets Google's Android which is used to run the majority of the world's smartphones including Samsung, HTC and Sony. WikiLeaks states that "These techniques permit the CIA to bypass the encryption of WhatsApp, Signal, Telegram, Wiebo, Confide, and Cloackman by hacking the 'smart' phones that they run on and collecting audio and message traffic before encryption is applied."

The CIA also runs a very substantial effort to infect and control

Microsoft Windows users. Many of these activities are pulled together by its Automated Implant Branch (AIB) which has developed several attack systems for automated infestation and control of CIA malware such as 'Assassin" and "Medusa."

Attacks against Internet infrastructure and web servers are the specialty of the CIA's Network Devices Branch (NDB).[182]

At least one US-based company, the cybersecurity firm Symantec, had been taking note and reporting on Vault 7 exploits for at least the prior six years. Symantec attributed the operations to a group named "Longhorn" which had applied techniques observed to almost exactly match technical details the organization used in cyber espionage activities targeted on organizations in more than sixteen countries.

Symantec investigators surmised that the CIA had been exploiting flaws in US-manufactured software for years without informing the companies involved. WikiLeaks alleged that this supposition was true; that a CIA program code-named "UMBRAGE" applied hacking tools that it had stolen from other governments to leave a misleading trail and cause investigators to believe attacks done by the CIA were, in fact, done by others.

In August 2017, a CIA employee named Joshua Schulte was arrested by the FBI. Schulte was charged with having passed over eight thousand pages of highly classified information to Julian Assange who had taken refuge in Ecuador's London embassy.[183]

In February 2019, NSA contractor Harold T. Martin III was convicted on multiple counts of mishandling classified U.S. government information. Martin had seized some 50,000 gigabytes of data obtained from classified NSA and CIA programs, along with source codes for numerous hacking tools.

By 2016, the CIA's hacking division had over 5,000 registered users and had produced more than a thousand hacking systems and malware utilizing more software code than used to run Facebook.

WikiLeaks editor Julian Assange stated that:

> *There is an extreme proliferation risk in the development of cyber 'weapons'. Comparisons can be drawn between the uncontrolled proliferation of such 'weapons', which results from the inability to contain them combined with their high market value, and the global arms trade.*[184]

The Colossal Capital One Caper

On July 19, 2019, Capital One Financial Corp., the fifth-largest U.S. credit card issuer, reported that a hacker accessed the personal information of approximately 106 million card customers and applicants, one of the largest-ever data breaches of a large bank.

As later confessed, a 33-year-old culprit, Paige Thompson, had broken through a capital One firewall to access data stored on Amazon.com Inc.'s cloud service between 2005 and early 2019. Thompson was a former employee at Amazon.com Inc.'s cloud division responsible for running much of Capital One's information-technology infrastructure.[185]

Thompson's breach compromised approximately 140,000 Social Security numbers and 80,000 bank account numbers, as well as some customers' credit scores, payment histories, and credit limits. She had intended for the data to be distributed online under the username "erratic."

The heist stands out not only as a massive breach but as a rare instance in which a former employee of Amazon has been charged with hacking the company's customers.

A previous 2017 outsider hack attack on the credit -reporting company Equifax Inc. had exposed the data of nearly 150 million Americans. From an identity theft perspective, the Capital One breach was less widespread than the Equifax hack because fewer Social Security numbers were compromised. Someone having another's Social Security Number enables them to more easily spin up an unauthorized account.[186]

Credit company hacking heists have a considerable history:

- 1984: TRW Information Systems and Sears (hit 90 million individuals).

- 2003: Data Processors International (hit 12 million credit card accounts).

- 2005: CardSystems Solutions (hit 40 million credit card accounts).

- 2009: Heartland Payment Systems (hit 130 million customers).

- 2014: JPMorgan Chase (hit 76 million households).

- 2014: Korea Credit Bureau (hit 104 million credit card accounts).

- 2017: Equifax (hit 148 million credit accounts).

- 2019: Capital One (hit 106 million credit card customers and applicants).

Giant corporate breaches typically have been the work of criminal teams, sometimes with ties to national governments. Thompson, on the other hand, was a self-described emotionally troubled individual who openly acknowledged her illegal activities in an online discussion group hosted by the social media collaboration company Slack Inc.

Social media posts from Thompson's Twitter account varied between mourning the loss of her cat—to being unemployed and experiencing homelessness—to struggles of being a transgender woman.

Thompson had changed her name in 2000 from Trevor Allen Thompson.

In one tweet just weeks before her arrest said she planned to check herself into a mental health facility.

Thompson's postings exhibited strange behavior of a deeply disturbed young person who both fretted and bragged about her exploits.

One Twitter message she posted under her erratic handle said:

> *I've basically strapped myself with a bomb vest, f*cking dropping capitol* [sic] *ones dox and admitting it."*

Another tweet said:

> *I'd give it at least two* [weeks] *before they find out who I am and the whole internet demands that I be banned.*

Thompson used a virtual private network and an anonymous web browser called "Tor" to shield her true hacking identity, first to gain access to Capital One's firewall, and then to access the bank's data on Amazon customers.

Steven Masada, assistant U.S. attorney for the Western District of Washington, told the Wall Street Journal that while Thompson's behavior might seem strange to outsiders—allegedly taking steps to conceal her identity while hacking Capital One, and then talking about her exploits publicly—is "not that uncommon in the hacker community that individuals brag about their accomplishments to seek recognition from their peers."[187]

Thompson's emotional insecurity seems inevitably tied to her spotty work and performance history. According to Amazon's internal ranking system her "Level 4" status was considered to be only a junior employee.

Before Amazon, Thompson had been jumping from job to job. Her resume listed three-month to two-year stints in engineering at Onvia, the now-closed Zion Preparatory Academy in Seattle and Acronym Media Inc., among other employers.

Acronym confirmed to the Wall Street Journal that Thompson worked remotely for the digital marketing agency for two and one-half months in 2011 and was terminated for poor work quality.[188]

Poor employee hiring and quality control wasn't Capital One's only information security vulnerability. People familiar with the penetration observe that the data stored on Amazon.com Inc.'s cloud involved a poorly constructed firewall—a mechanism designed to wall off privately operated digital systems.[189]

Cloud computing use boomed as companies turned to providers such as Amazon and Microsoft Corp. to do the work of configuring computers inside their own data centers. The processing power of those servers and storage devices is then rented out to customers who pay, depending on how much work the computers do.

Many financial institutions gradually moved customer data out of their data centers as more and more major global companies such as JPMorgan Chase & Co. and Bank of America Corp. became members. This cloud computing trend caught on in part because it allowed software engineers to sidestep cumbersome security restrictions and sluggish development processes that made companies' in-house technologies clunky.[190]

As the Capital One hack demonstrates, the ease and speed of opting instead to fire up a server through Amazon Web Services also led to many cloud misconfiguration problems that can leave sensitive data exposed to unauthorized access with catastrophic results.

Vincent Liu, a partner with the security consulting firm Bishop Fox, observes that "basic cyber hygiene gets thrown out the window" as companies move to new technologies such as the cloud.[191]

NSA Snoops on the Google Cloud

Former CIA and Booz Allen Hamilton employee Edward Snowden's trove of leaked documents included a hand-drawn sketch diagram on a yellow piece of paper with ominous privacy intrusion implications for everyone connected to cloud servers. First released by the

Washington Post on October 30, 2013, a NSA presentation slide marked *"TOP SECRET//SI//NOFORN"* revealed that NSA was trying, perhaps successfully, to insert itself in the nexus between the public Internet and the "Google Cloud."

By finding a way between the two servers—"man in the middle" attack—the NSA would be able to intercept all kinds of traffic moving between them and the outside world, from gmail messages to Google map searches.

While not specifying exactly how the NSA was planning on getting between those servers, David Sanger postulates that NSA might physically tap into undersea cables from the U.S. or allied base where they come ashore. Since Google had not gotten around to encrypting the data that was "in transit' through these cables, merely getting into the network itself was the price of admission to the data.[192]

The NSA drawing included an arrow pointing only to the schematic place in the diagram that corresponded to where NSA was inserting itself. A smiley face that the author had doodled next to the arrow not only added insult to injury to Google, but presented a stark warning to Facebook, Apple, and every other major information platform and service company.[193]

They had good reasons for alarm. If Google was being hacked by the U.S. government, any information leaks could be attributed to them as complicit entities. Customers in Germany or Japan, for example, would have good reason to suspect that they or any associated American companies were secretly turning over their data to the NSA.

Foreign actors and entities were certainly not the only ones warranting concern. NSA's digital spycraft would enable access to many millions of American accounts as well. With the advent of smartphones, the public was coming aware for the first time that they were keeping their whole lives in their pockets. Most all of their medical data, their banking information and work emails, their texts with spouses, lovers, and friends are being stored in Google servers, and others like it run by Yahoo!, Microsoft, and smaller competitors.

If served with a private information order from the Foreign Surveillance Court, Google and other big communications providers have little to no choice but to comply. About data harvesting on "US Persons," the NSA would merely have to get a friendly court order. Anyone who wasn't legally an "American person" would be a fair game with no court orders required.

Also, depending on where one was at the time, their data could be stored anywhere. This virtually wiped out any distinction between "international," versus" domestic," communications. All of a sudden the idea of government getting inside Google's servers seemed a lot more worrisome. And although born an American company, Google saw itself as much more—a global citizen, a company responsible for global customers.

Google's official response to the NSA slide discovery was immediate and predictable:

> *We are outraged at the lengths to which the government seems to have gone to intercept data from our private fiber networks, and it underscores the need for urgent reform.*[194]

Google also soon added a new mail encryption feature to its product line. As a pointed jab at NSA, the code ended with a smiley face.

For the first time in post-World War II history, American firms at least publicly refused to cooperate with their government. Some of this refusal can be logically attributed to Silicon Valley's famously libertarian ideology. Other motivations are attached to the legitimate fear that any open association with the NSA will prompt their customers to wonder whether Washington had bored holes in their products.

David Sanger observes that the Snowden revelations, in general, set the stage for massive conflicts between the powerhouses in the technology world—Google, Apple, Facebook, and Microsoft—and a National Security Agency that had blithely assumed American companies would be on its side.

Other revelations fueling that heated contention had surfaced even before Google's smiley face emerged. These WikiLeaks had revealed that Google and its servers were but a small part of a far broader NSA domestic spying effort.

PRISM Shines Light on NSA Overreach

A June 7th 2013, *Guardian* article revealed the existence of the NSA program named "PRISM," in which NSA siphoned off Internet communications of all types under orders issued by the Foreign Intelligence Surveillance Court. The companies were ordered to stay quiet about the program and were paid several million dollars in compensation for compliance costs.[195]

Most, or all of them, from Microsoft—to Yahoo!—to Apple—to Skype went along with the demands and bribes, although some did so more actively than others.

Mark Zuckerberg posted a heated defense:

> *Facebook is not and has never been part of any program to give the U.S. or any other government direct access to our servers.*

On the same day, Microsoft declared on that same day that it complied "only with orders for requests about specific accounts or identifiers." They added:

> *If the government has a broader voluntary national security program to gather customer data, we don't participate in it.*[196]

David Sanger noted the irony in the fact that as tech firms publicly protested the NSA's intrusions into their networks—Microsoft,

IBM, and AT&T among them—they were privately vying for the hugely profitable business of managing the intelligence community's data.

The Snowden WikiLeaks documents exposed a long and deep history of sweeping surveillance collaboration between NSA and telecommunication giants that had become the backbone of the Internet that global public and private sector customers rely upon.

A large trove of evidence revealed that seventeen AT&T Internet hubs in the United States had installed NSA surveillance equipment and; a smaller number of Verizon facilities were similarly equipped.

Sanger comments:

> *It was a relationship so vital that President Obama sometimes intervened directly with telecommunications executives, calling them personally to ask for help if there was a critical need for information about a terrorist group or if intelligence agencies feared that an attack could be imminent.*[197]

Silicon Valley companies broadly agreed that institutional banking information and certainly classified government files on the cloud should be encrypted, yet the idea of doing the same regarding every individual's personal data was relatively new.

Sanger observes that although the Pentagon began moving in the direction of building its cloud storage system, the whole idea of centralizing information in the cloud made many senior Army officials exceedingly nervous. Whereas the 9/11 tragedy demonstrated a need for more access to shared security information, the Snowden lesson revealed that centralized systems can yield huge security leaks and other threats.

Recent history has made it abundantly clear that globally-wired-together Internet and cloud storage present epic game-

117

changing public and private security challenges. This vast new cyberspace warfare front is unlike previous conflict domains dominated exclusively by the most powerful nations or highly funded and advanced weaponry geniuses.

The Internet age has opened up an online flea market for every sort of infectious and destructive bug that all those with the money can afford.

Besides, unlike traditional warfare conflicts that began when openly declared and end when strategic objectives are achieved or abandoned, cyberwarfare is covert, perpetual, and ultimately, unwinnable. By its very nature, cyber defenses lag ever-increasingly wired-together vulnerabilities, viruses, victims, and villains.

Although the United States likely possesses the world's most sophisticated offensive cyberwar capabilities, that offensive prowess cannot make up for an inherent weakness in our defensive position. Ironically, as former U.S. Admiral McConnell has noted:

> *Because we are the most developed technologically—we have the most bandwidth running through our society and are more dependent on that bandwidth—we are the most vulnerable.*[198]

CYBERSURVEILLANCE TOOLS FOR TYRANTS

THE INTERNET IS a huge game-changer concerning how we, as individuals and society, assess our enemies, meet their challenges, and enact policies that impose and limit intrusive threats to everyone connected together in a globally wired-together information and infrastructure world controlled by enormously powerful and influential organizations.

As former George G.W Bush administration, Special Assistant for Homeland Security Marie O'Neill Sciarrone warns, this expansionary nature of the Internet exceeds all geographical boundaries and the reach of international norms—including the rules of the Geneva Convention.[199]

Those affected by the ever-expanding Internet and AI-enabled cyberwarfare threats are no longer just major public and corporate entities. Countless millions of everyday citizens have become direct targets as well as collateral damage, bringing the cyberwarfare battlefield into our homes and personal lives.

119

Global adversaries not only have capabilities to hold critical infrastructures at risk but also threaten the far broader ecosystem of connected consumer and industrial devices known as the "Internet of Things (IoT). Included are connected residential and business thermostats, cameras, televisions and even cookers that can all be used either to spy on complacent, unaware citizens.

I briefly introduce these threats in my most recent 2019 book: *The Weaponization of AI and the Internet*:

> *Human society is experiencing the very earliest beginnings of an expansively transformative and disruptive information revolution.*
>
> *Exercised through the minds, hands and free choices of the many, such benefits are endless. Controlled by the special interests and agendas of but a few, the applications for social control and exploitation are equally boundless.*
>
> *Remote surveillance and monitoring technologies follow and record U.S. virtually everywhere—even in our homes and cars.*
>
> *Ever-present smart phone "voice assistants" listen in on our private conversations and snitch on U.S. to uninvited outsiders.*
>
> *Electronic eavesdroppers catalog our special interests, analyze our psychological profiles, and micro-target U.S. for messaging.*
>
> *Personal viewpoints expressed in social media exchanges are monitored and censored by politically-biased algorithms. Expansively wired-together Internet of Things networks and a rapid emergence of "smart cities" openly invite autocratic control.*
>
> *Be very cautious of all-too-seductive invitations to trade away precious privacy for promises of*

increased convenience, efficiency, and security from predators.

Consider the fate of a frog in a shallow pan of water placed over a flame complacently adjusting the temperature change until it's too late to jump out.[200]

Serious Games of Chinese Checkers

We in America have connected more of our economy to the Internet than any other nation. Of the eighteen civilian infrastructure sectors identified as critical by the Department of Homeland Security, all have grown reliant on the Internet to carry out their basic functions, and all are vulnerable to cyberattacks by nation-state actors.[201]

These circumstances starkly contrast those in China where the networks that make up their Internet infrastructure are all controlled by the government through direct ownership or very close partnership with the private sector.

In design and practice, China's Internet system operates more like the internal network of and an independent company—an "intranet"—where the government functions as a service provider, content controller, and security protector. The network is segmented between government, academic, and commercial uses, all of which operate under the domain of government control.

Segmentation in its design and isolation by its "Great Firewall" afford China both the discretionary power and means to disconnect its connected slice of the Internet from the rest of the world. This combined capability enables authoritarian operators to block user access to disapproved outside information; to exercise population control through technologies that monitor behaviors and screen electronic communications deemed illegal and; to facilitate network protections against foreign hackers and malware.

Popular websites such as Google, YouTube, Facebook, Twitter and Wikipedia too, are banned in China in the Name of "national

121

security." These policies also result in securing Beijing from challenges to Communist Party rule through the free flow of two-way outside information and communications.[202]

In China, it is the government's role to actively defend the network and its users. Not so in the United States, where the government has no such overall authority or capability because its connections to the Internet are privately owned and operated.

Similarly, China, and North Korea also, have offensive and defensive cyberattack advantages afforded by abilities to disconnect critical from nonessential network functions. "Defense" in this cyberwar context, is a measure of a nation's ability to take actions that under attack, which will block or mitigate the offense.

Again, by contrast, U.S. military cyber defenses depend upon a private system of unified and integrated service providers.

China has been developing its own proprietary operating system that would not be susceptible to existing network attacks along with computer software allegedly meant to block children from gaining access to pornography. The most likely real intent is to give the government control over every desktop in the country.

Among the first to confront the problem was Google, whose experience taught every American company that China wasn't hacking just for hacking's sake: it had an intelligence angle and a political agenda.[203]

As American intelligence agencies later learned, an unfettered Google had made Chinese politicos very nervous. Leaders who Googled themselves often weren't always pleased with what they found.

A secret U.S. State Department cable obtained in the WikiLeaks trove revealed that Li Changchun, who headed the propaganda department of the Chinese Communist Party was astounded to discover in a Google "results critical to him." [204]

Selfie search results weren't Beijing's only Google issues. Li also commented that he didn't like Google Earth, the satellite mapping software that displayed "images of China's military, nuclear, space, energy, and other sensitive government agency installations."

Li demanded that state-owned Chinese telecommunications firms cut off Google's outreach to hundreds of millions of Chinese users unless it complied with China's censorship rules.[205]

In mid-December 2009, Google corporate officials discovered that the company had been targeted by a highly sophisticated Chinese cyber espionage campaign.

In addition to intellectual property, the hackers were looking for broad forms of political and strategic intelligence. They had raided Chinese dissident email accounts, collected information about the activities of Chinese nationals living in the United States.

The hackers also monitored email communications of key American decision-makers who used Gmail because was unsecured from government computers. They even mapped where the targets worked and identified their special strategic vulnerabilities.[206]

Google's hackers applied the same cyber-snooping tradecraft including spear-phishing spoofs that are broadly applied to far less technically savvy organizations.

A Google executive or senior research scientist, for example, might receive an email containing a link to a website that appeared to be from a colleague. The message might say, "Hey, Joe, I think this story will interest you..." and then provide a link to a fairly innocuous site.[207]

Visiting the site infected the spoofed victim's computer with malware buried in corners of Google's corporate networks where it could easily be missed. Once it was dug in, the malware created a covert back door communications channel reaching throughout the company network to reach the servers containing the all-important secret source code.[208]

Google scientists first traced the hacking back to a server in Taiwan where they found copies not only of their proprietary information but thefts from thirty-five other companies as well. Included, were hacked materials from Adobe, Dow Chemical, and the defense contractor Northrop Grumman.

Dubbed by investigators as "Operation Aurora", the cyberattack trail ultimately led to Mainland China. Google then notified the FBI and shut down its Beijing operations in 2010.[209]

Other major American information tech companies, including Facebook, Apple Uber, Apple, and Microsoft, encountered Chinese business rules that leave similarly stark options: either turn over their company's information, and often their underlying technology, or like Google... get out.

China's Microsoft strategy leveraged the company's own software vulnerabilities against them.

Microsoft Windows—one of the world's most popular operating systems—was also long-recognized by hackers as having the most badly-written, or "buggy" computer software code. The Windows program served as the operating system for desktop and laptop computers, and also for Cisco's Internet routers.

Both Windows systems were proprietary, therefore not publicly available. Although the software could be purchased as a finished product, the underlying code was held a secret.

China which represented a huge Windows market threatened to develop its own competing Linux system it called "Red Flag" unless Microsoft dropped their price and gave them their secret code. Microsoft caved into the blackmail demand and even established a software research lab in Beijing that was wired to their U.S. headquarters.[210]

China's "innovation through plagiarism" achievements have been enormously prolific.

A study by Verizon's Data Breach Investigations team found that "96 percent of recorded, state-affiliated attacks targeting businesses' trade secrets and other intellectual property in 2012 could be traced back to Chinese hackers." [211]

We should not, however, count on China to languish and lag behind other countries in future proactive hi-tech competitiveness. According to a 2019 report from the Council on Foreign Relations (CFR), China is vastly outpacing the U.S. is planning for and investing in critical research and producing more top minds in AI and advanced quantum computing. By the year 2030, China will be the world's largest spender in these areas.

Adam Segal, an emerging technologist and national security expert at CFR, reports that "China is producing three times as many science, technology, engineering and mathematics (STEM) graduates at the undergraduate level" as the US.[212]

Over the last generation, China's economic boom over the last generation was jump-started by the production of cheap goods using foreign intellectual property. Government leaders clearly recognize that dependence on this approach will not serve future economic needs. A Chinese factory that made early model iPhones, for example, earned only about $15 per phone for assembling a $630 iPhone.[213]

A program established by the George W. Bush administration to defend the U.S. government against cyberwars—a previously secret "Supply Chain Initiative" (PDD-54), mandated that the U.S. should only purchase software and hardware made and purchased in America under secure conditions. That solution, however, is far from simple to achieve.

Chips, the guts of all common computers can be customized and compromised, just like their software. Most experts cannot look at a complicated computer chip and determine whether there is an extra piece here or there, a physical trap door.

Those chips were originally made in the US. Most are now manufactured in Asia.

China's overbroad national security restrictions create an additional irony for Huawei. President Trump has blacklisted Huawei—a company that has built a huge overseas business using American technology—for suspected espionage.

Huawei's smartphones are popular in markets outside China. That demand will slump sharply if Trump bans access to Google's ecosystems and apps.

Larry Bell

The Ugly Side of Facial Recognition

Some of China's most advanced information technologies are being implemented as weapons of mass detection and deception against its citizens.

A vast, ubiquitous AI-enabled facial recognition network will reportedly be capable of picking any one of China's 1.4 billion individual citizens out of massive crowds with nearly perfect accuracy. Their state-run Xinhua news agency reported that Nanchang police in southeastern China located and arrested a wanted suspect at a 60,000-person pop concert event.

US agencies, of course, possess at least comparable capabilities. For example, AI facial recognition technology is known to have been used by the U.S. Department of Homeland Security, San Diego's Police department and others to enhance security at large events like the Super Bowl.

In the hands of autocrats, however, the technology has great potential for repressive use by strongmen and police states to bolster their internal grip, undermine basic rights, and spread illiberal practices beyond their own borders.[214]

Chinese police deployed facial recognition glasses in early 2018, and Beijing-based LLVision Technology Co. sells basic versions to countries in Africa and Europe. Such glasses can be used to help identify criminals like thieves and drug dealers—or to hunt human rights activists and pro-democracy protesters.

Known in China as "Pingan Chengshi" (or Safe Cities), an estimated 170 million "Sky Net" cameras were installed across China in 2018, with another 400 million to follow. When installations are completed, "Beijing plans to be able to identify anyone, anytime, anywhere in China within three seconds."[215]

In combination with an expansive facial recognition and social media surveillance initiative, China will implement a "social credit system" which will assign every citizen merits and demerits according to how closely they are observed to conform to official behavior standards. Although initially voluntary, the program will become mandatory in 2020.

Examples of demerit-penalized infractions are expected to include bad driving, smoking in non-smoking zones, purchasing too many video games, paying bills late, breaking family planning rules, posting incorrect information online, jaywalking, or walking a dog unleashed. Those with low scores will be blocked from booking domestic flight and luxury business-class train tickets, staying at premium hotels, purchasing real estate, getting hired, having children denied acceptance to university programs, and even risk having their pets confiscated.

According to the plan issued by Beijing's municipal government, by 2021, the capital's blacklisted citizens will be "unable to move even a single step." [216]

Chinese authorities are a priority- targeting and testing the powerful tools of facial recognition and big data to detect departures from "normal" behavior among ethnic Uighurs and other Muslims in their Xinjiang region, a swath of desert and mountains abutting Central Asia. Many among the roughly 12 million Uighurs, mostly a Turkic population, are being imprisoned in political indoctrination camps with purported aims of assimilating them with the country's Han Chinese majority. [217]

Xinjiang detention centers reportedly resemble those established through China's now-defunct "laojiao", or "re-education through labor," system that herded criminals and dissidents into work camps before it was abolished in 2013.

Represented by officials as "vocational training centers," more than a million detainees in the new camps—chiefly Uighurs—have been forced to watch videos about President XI Jinping and the Communist Party, sing patriotic songs, and denounce Islam.

China has also embedded more than a million party members and government workers, mostly ethnic Han Chinese, to live with Uighurs and other minorities for week-long stays in their village temples and homes to monitor individuals and families to recommend whom to detain.

Besides, Xinjiang authorities have collected biometric data, including blood samples, from all residents between ages 12 and 65, turning the region of 24 million people into a leader in efforts to build a national DNA database.

Although this repressive treatment of Uighurs has drawn an international outcry, some of the same totalitarian crackdowns have filtered to other parts of the country amid efforts to reassert Communist Party dominance overall Chinese society.

Some of the oppressive technologies and tactics applied in Xinjiang were first implemented in other provincial regions including Tibet.

Thousands of Tibetans deemed influenced by the Dalai Lama, whom Beijing denounced as a separatist, were also forced into re-education camps. Many others had their passports confiscated. Exiled Tibetan filmmaker Dhondup Wangchen who had completed a six-year prison term on a subversion charge characterized his country as having become "the world's largest prison"—one blanketed with checkpoints and security cameras.[218]

Building and expanding upon that Tibetan achievement, the Xinjiang program created a police state that created a police regime previously unmatched both in scale and sophistication.

Many thousands of high-tech police stations throughout Xinjiang tapped big data analytics to collect and sift through vast pools of personal information—such as an individual's movements, banking, health, and legal records—to identify potential issues of concern.

Security forces were greatly expanded with orders to "fight till the terrorists are stricken with fear." Police were availed with hand-held devices to scan photos, messages and other data in residents' mobile phones in searches for sensitive materials.

Expansive networks of security cameras were linked with police databases.

Thousands of small urban "mini-community centers" spaced

about 1,000 feet to 1,600 feet apart provided amenities including household tools, first-aid kits, and medications, wireless internet, phone chargers—along with covert devices to extract data from those mobile phones.

Chinese officials are actively promoting Xinjiang as a model for developing countries. The program director has reportedly hosted more than 200 politicians from nearly 30 countries, including Russia, Egypt, and Turkey, for a symposium on Xinjiang's ethnic policies.[219]

A political dissident in Harare, Zimbabwe may soon have as much to fear as a heroin smuggler in Zhengzhou. The Chinese AI firm Cloudwalk Technology has sold Zimbabwe's government a mass facial recognition system that will send data on millions of Zimbabweans back to the company in China. Cloudwalk will then refine its algorithms and perfect the system for global commercial export.

Business is also booming for other Chinese surveillance technology companies. The global client list of one such firm Tiandy, a CCTV camera manufacturer and "smart security solution provider," includes more than 60 countries.

As reported in *The Washington Examiner*, China is already distributing those same intrusive tools of surveillance and social control throughout Latin America with potential threats "surrounding the United States".

Quoting Colorado Senator Cory Gardner, "China's goal is to displace the United States, and they can do that by wreaking havoc in the Western Hemisphere."[220]

According to the U.S. and regional officials, ZTE Corp, China's second largest telecommunications equipment maker, has played a key role in these threats, both in their country and abroad.

ZTE is known to have been helping the Iranian regime track dissidents since at least 2010.

Venezuelan strongman Nicolás Maduro's regime paid the company $70 million to create a database and payment system for a "homeland card" (carnet de la Patria) for a social control program that is modeled after China's social credit system. The card which is used to control access to food, cash bonuses, and other social services also serves as a political control mechanism, recording where users gained access to food and medicines in a country with shortages of both.

Human Rights Watch reports that the homeland card may capture voting history as well. According to a 2028 investigation by *Reuters*, the data that this system generates is stored by ZTE, which has reportedly deployed a team of experts within Venezuela's state-run telecommunications company Cantvu to help run the program.

China has also supplied other governments in the region with substantial surveillance capabilities that can be used for social control. Ecuadorian law enforcement officials have purchased a network of Chinese security cameras with facial recognition software.

Evan Ellis, a Latin America analyst at the Center for Strategic and International Studies and professor at the Strategic Studies Institute of the U.S. Army War College warning that If enough countries adopt Chinese technology, Beijing will be able to integrate the different assets into a transnational' "incredibly effective surveillance complex" that reinforces China's political and economic influence.[221]

This is already occurring.

The Egyptian government plans to relocate from Cairo later this year to a still-unnamed new capital that will have, as the project's spokesman put it, "cameras and sensors everywhere," with "a command center to control the entire city."

Moscow already has some 5.000 cameras installed with facial recognition technology, and it can match faces of interest to the Russian state to photos from passport databases, police files and even VK, the country's most popular social media platform.[222]

Meanwhile, China plans also plans to upgrade more of its smart cities. Like Yinchuan, for example, where commuters can use a positive facial ID to board a bus, and Hangzhou, where facial data can be used to buy Kentucky Fried Chicken meal.

And they're really only just getting started. Planned megacities like Xiongan New Area, a development southwest of Beijing, suggests the shape of future "panopticons". These "smart cities" will feature centralized systems of control across financial, criminal and government records, drawing on all websites, visual imagery, phone applications, and sensors.

Here, as in China and other countries, spy cameras are sprouting up on lampposts and rooftops everywhere. Facial recognition systems can detect and record each of our individual movements. Our automobile and cellphone locations are constantly recorded. Our online shopping and lifestyle interests are being tabulated and distributed to product and service agencies. Our smartphones can constantly eavesdrop and secretly pass along private information.

There's no need to travel abroad to have your face unknowingly appear on candid camera. New York's Metropolitan Transportation Authority has begun installing monitors to scan motorists' faces at bridges and tunnels connecting Manhattan to other boroughs. Five seconds after a car enters the crossing, its drivers face is processed and compared to a state database. Police on the other side is waiting to nab people for warrants, suspected felons, parole violators, and terrorist suspects.

As Gov. Andrew Como added, that's not all:

> *Because many times a person will turn their head when they see a security camera, they are now experimenting with technology that just identifies a person by their ear, believe it or not.*[223]

George Orwell's grim 1984 story admonition that "Big Brother is Watching You" has gained a rapidly growing number of progeny years he wrote it in 1949. That same year an American company released the first commercially-available closed-circuit television (CCTV) system.

The market for these surveillance devices exploded a half-century later following the September 2001 terror attacks which shook the national psyche. By 2018, New York City had ramped up installations to roughly 20,000 officially-run cameras in Manhattan alone. Chicago meanwhile had installed an estimated 32,000 CCTVs to help combat the inner-city murder epidemic.

President Trump signed an executive order in 2017 calling for facial biometric scanning of all international travelers through America's top 20 airports by 2021. The Department of Homeland Security is assembling a biometric database called Homeland Advanced Recognition Technology, and the National Security Agency scans and stores millions of faces on the web. Facebook, among many other social media companies, collects images of its users' faces.

As reported in the Hong Kong-based *South China Morning Post* (SCMP) owned by China's large Alibaba Group conglomerate, the New York City Police Department (UHPD) is using the same facial recognition surveillance tools developed in China to monitor it, American residents.

According to a January 11, 2019 article, "Hikvision—the world's largest surveillance technology, which is state-owned and based in Hangzhou in eastern China—has supplied the equipment and software used by an American force that polices a population of about 8.6 million people." [224]

Whereas the full extent and use of Chinese cameras and software in the NYPD network was unknown, a reported official document described it as being on a "large scale".

SCMP noted that Hikvision's website has shown that thousands of their artificial intelligence-aided Sky Net cameras have been monitoring River Park Towers in the Bronx—one of New York City's largest government-subsidized low-income housing developments—since 2014.

Although well-known for a checked history of arson, drug rings, and gang violence, police responders have been faced with accusations of racially-biased criminal profiling of African-Americans in combination with heavy-handed stop-and-search tactics and home searches without warrants.

Hikvision's cameras have been installed at River Park Towers entrances, lobbies, corridors and stairwells in an effort to defeat vandalism. According to an article in *Security Products Magazine*, NYPD shares direct access to the network with the property's management company.[225]

Hikvision's surveillance system is claimed to be far more accurate than competing Western technologies in identifying faces of different racial complexions. The software can also spot and follow a person according to their body, posture and walking characteristics.

A researcher for the National Laboratory of Pattern Recognition at the Chinese Academy of Sciences in Beijing told SCMP that machine recognition of faces with dark skin tones had been a major challenge for artificial intelligence developers for years, requiring "leaps and bounds" of technology.

A 2018 MIT study found that while a Microsoft machine could recognize white males every time, there was a more than 20 percent margin of error when used to identify a black woman.[226]

Chinese developers at Hikvision created an algorithm which overcame unclear and distorted low-light images to enable their cameras to identify distinctive facial features at night with "just a few stars blinking in the sky." [227]

The SCMP reported that Hikvision products and technologies have also been used across the US, including along 2,369 miles of U.S. Route 1, America's longest north-to-south road which connects major east coast cities.

The Internet of Things and Threats

Legitimate personal and public privacy and security concerns arise where just about every electronic device now has some kind of shared connection. Examples are smart meters, thermostats, smart speakers, web cameras, fitness trackers, health care monitors, and many kinds of child toys.[228]

The Internet-of-Things involves adding internet connectivity to a system of interrelated computing devices, mechanical and digital machines, objects, animals and/or people. Each "thing" is provided a unique identifier and the ability to automatically transfer data over a network. Allowing devices to interconnect opens them up to a number of serious vulnerabilities.

The Federal Trade Commission's National Telecommunications and Information Administration, a unit of Commerce warned that while there are many IoT benefits for customers, "these devices also create new opportunities for unauthorized persons to exploit vulnerabilities." [229]

Many IoT hacks don't target the devices themselves, but rather use IoT devices to access laptops and other systems holding such personal information as name, age, gender, email address, home address, phone number, Social Security number, and banking and account and social media accounts.[230]

Security experts have long warned of the potential risk of large numbers of unsecured devices connected to the Internet since the IoT concept first originated in the late 1990s. A number of attacks have made headlines, from refrigerators and TVs being used to send spam to hackers infiltrating baby monitors and talking to children.

One of the key security problems noted is the impracticability of updating them when vulnerabilities are discovered. Installing new firmware on light bulbs or refrigerators is not something most consumers are used to, and many manufacturers haven't contemplated those processes either. This presents a serious vulnerability for consumers and businesses alike.

The National Telecommunications and Information Administration reports that although similar risks exist with traditional computers and computer networks, they may be heightened in the IoT, in part because many IoT chips are inexpensive and disposable, and many IoT devices are quickly replaceable with newer versions:

> *As a result, businesses may not have an incentive to support software updates for the full useful life of these devices, potentially leaving customers with vulnerable devices. Moreover, it may be difficult or impossible to apply updates to certain devices.*[231]

Common hacking targets include attacks against connected medical devices, home monitoring equipment, and industrial control systems which come with hard-coded passwords or credentials that can be discovered quite easily. When breached, they often lead to larger home or business networks, offering attackers easy footholds for further attacks.

Adding to privacy risks, many connected devices collect information about their usage, environments, and users, which is then used by vendors for any number of purposes. One FTC analysis discovered the presence of numerous third parties in apps connected to IoT health and fitness wearable devices which revealed medical search histories, along with zip code, gender, and geolocation data. The FTC report said:

> *The massive volume of granular data collected by IoT devices enables those with access to data to perform analyses that would not be possible with less rich data sets.*

As discussed in my previous book *The Weaponization of AI and the Internet,* those same wired-together Internet-of-Things personal and public devices that are supposed to make our lives safer and more efficient, and those who manage them, will be fully capable of tracking, scrutinizing, and ultimately controlling everyone—at every place—and at any time.[232]

The grand-scale IoT concept of "smart cities" became an American buzzword long before the advent of the Internet. The term arguably dates back to the first implementation of automated traffic lights in 1922 in Houston. The idea has since morphed and crystallized into an image of "the city" as a vast, efficient robot—a dream of giant technology companies.

Advancing digital information communication technologies (ICT) are now transforming massive AI-coordinated data sets into modern smart city planning tools, policy guidelines, and management systems. These applications follow two alternate technological visions.

The first and most prevalent vision perceives the urban fabric and everyone in it as becoming increasingly instrumented and ubiquitously monitored and controlled from "everywhere." Here, ICT is typically viewed as a primary driver of change rather than being relegated to only serving as but one means to move the cities higher on the development ladder. In this new language of "smartness," the emphasis is upon smart information, smart meters, smart grids and smart buildings which are rapidly becoming ever-larger parts of our everyday lives.

The second perception envisions the smart city's economy driven by innovation and entrepreneurship with the goal of attracting business and jobs and focusing on efficiency, savings, productivity, and competitiveness. ICT's role here is to facilitate and streamline private and public initiatives. The idea of smartness here focuses on smart use of resources, smart and effective management, and smart social inclusion. Within this view, the ICTs are one component of the concept, but by no means its bread and butter.

There is a dichotomy in this picture: corporate utopian visions ("ICT will save us") vs. an academic circle definition that is more varied, diverse and complex. Nevertheless, the two perspectives have morphed to become applied interchangeably, an ambiguous and not always positively connoted mix of automated utopian fantasy lands overseen by robotic versions of George Orwell's Big Brother.

"Smart cities" have come to be marketed as a conveniently imprecise, rhetorical and ideological rationalism for technology reigning supreme over everything... a panacea for all urban ills. Promoters market ICT solutions as quick and effective ways to deal with all manner of urban problems: growing populations, climate change, environmental shocks and other urban threats, including local urban crime problems, congestion, inefficient services, and even economic stagnation.

Such rhetoric energetically promulgated by big technology, engineering, and consulting companies is predicated on the embedding of computerized sensors into the urban fabric so that bike racks and lamp posts, CCTV and traffic lights, remote-control air conditioning systems and home appliances all become interconnected into the wireless broadband Internet-of-Things.[233]

Whole new cities, such as Songdo in South Korea, have already been constructed according to this template. Songdo's buildings have automatic climate control and computerized access; its roads, water, waste and electricity systems are dense with electronic sensors to enable the city's brain to track and respond to the movement of residents.

In India, Prime Minister Narendra Modi has promised to build at least 100 smart cities. Dholera, the first, is currently an empty backdrop landscape-in-waiting due in large part to flooding problems that have discouraged private investment. Despite promotional arguments that the city was planned as a green-field development to serve the "public good," peasant and farmer protestors who lost their lands angrily disagree.[234]

One of the most ambitious pilot projects of this kind was in Rio de Janeiro and involved constructing a large state-of-the-art Center of Operations operated by IBM. Branded as "Smarter Planet," it featured impressive control center dashboards to provide panoptical views of huge amounts of data.

Rio made the great investment into the smart system as a tool to predict and manage a flood response in advance of two huge events—FIFA World Cup in 2014, and the Olympic Games in 2016. That purpose soon morphed into an enormously larger urban surveillance information gathering and processing enterprise. Quoting Rio's mayor, Eduardo Paes:

> *The operations center allows U.S. to have people looking into every corner of the city, 24 hours a day, seven days a week.*[235]

Writing in *The Guardian*, Steven Poole warns:

> *The things that enable that approach—a vast network of sensors amounting to millions of electronic ears, eyes and noses—also potentially enable the future city to be a vast arena of perfect and permanent surveillance by whomever has access to the data feeds.*[236]

Trolling for Our Personal Information

China and Russia are poised to use and export a new suite of AI psychological propaganda products and capabilities that will enable even second-tier tyrannies will be able to better monitor and mislead their populations.

One key technology in this in this arsenal applies automated personality assessments to tailor messaging slants that micro-target population segments based upon their particular psychological, demographic or behavioral characteristics.

In a widely-viewed TED Talk in 2017, techno-sociologist Zeynep Tufekci described a world where "people in power [use] these algorithms to quietly watch us, to judge U.S. and to nudge us, to predict and identify the troublemakers and the rebels."

This is enabling the tailoring of far more effective "influence campaigns" aimed both at domestic and foreign population targets.

According to a *Wall Street Journal* report, Russia's Internet Research Agency harvested data from Facebook to apply this capability during the U.S. presidential race. Data harvested from Facebook postings were used to craft specific messages for individual voters based in part on race, ethnicity and identity.[237]

Facebook and Google are known to have applied micro-targeting algorithms for precision advertising. Google, for instance, labeled users as "left-leaning" or "right-leaning" for assistance to Democrat political advertisers in the 2016 election.[238]

The more powerful micro-targeting becomes, the easier it will be for autocracies to influence speech and thought. As some American companies lead the way, more and more government entities—including tyrannical regimes—are certain to follow.

Speaking at an October 2018 discussion at the Council on Foreign Relations, former director of the U.S. government's Intelligence Advanced Research Projects Activity, Jason Matheny, cited "industrialization of propaganda" as a growing international threat. For this reason, he warned the U.S. intelligence services to beware of the "exuberance in China and Russia towards AI."[239]

AI-driven applications will soon also enable authorities to analyze patterns in a population's online activity to identify and target those who are most susceptible to a particular propaganda message.

Emerging technologies are also rapidly changing ways that autocrats deliver mass propaganda through the use of online "bots" (automated account messaging). For example, researchers at New York University found that fully half of the tweets from accounts that focused on Russian politics during their 2014 invasion of Crimea and in the months afterwards were bot-generated.

Similarly, the October 2018 murder of Washington Post columnist Jamal Khashoggi prompted a surge in messaging from pro-regime Saudi bots.[240]

Bot messages will soon be indistinguishable from humans online—capable of denouncing anti-regime activists, attacking rivals and amplifying state messaging on alarmingly lifelike ways. As Lisa-Marie Neudert, a researcher with Oxford's Computational Propaganda Project has warned:

> *[T]he next generation of bots is preparing for attack. This time around, political bots will leave repetitive, automated tasks behind and instead become intelligent.*[241]

Neudert told the attendees International Forum of Democratic Studies in October 2018, that the same kind of tech advances that fuel Amazon's Alexa and Apple's Siri, are also teaching bots how to talk. Speech-synthesis systems made by companies such as Lyrebird (which says it creates "the most realistic artificial voices in the world") require as little as one minute of original voice recording to generate seemingly authentic audio of the target speaker.[242]

The Chinese government has employed what's known as the "50 Cent Army" over many years when thousands of fake, paid commenters have posted online messages which favorable to Beijing to distract online critics. In the future, bots will do the work of the current legions of regime-paid desk workers.

Advanced generations of "deep fakes"—digital forgeries combining audio with video imagery are already becoming good enough to fool many listeners and viewers. On YouTube, one can already see an unnerving mashup of actors Steve Buscemi and Jennifer Lawrence and a far-from-perfect video made by the Chinese company iFlytek showing both Donald Trump and Barack Obama "speaking" in fluent Mandarin.

These increasingly convincing bots will inevitably work together with other new tools to let dictators spread chillingly insidious disinformation. Dartmouth computer science professor Hany Farid warns that explosive growth in the number of these weaponized fakeries will leave those playing defense "outgunned".

Farid estimates that there are probably 100 to 1,000 times

"more people developing the technology to manipulate content than there is to detect [it]... Suddenly, there'll be the ability to claim that anything is fake. And how are we going to believe anything.[243]

More than 30 billion devices are expected to be connected and generating data on the Internet by 2020. Any organization or insider that can control, process and exploit this information to spy on others in the network will be weaponized with enormous social and economic advantages.

Information harvesting and processing is being greatly enhanced by rapid advances in AI machine learning and data storage capacities. Added to this, companies in both the U.S. and China are optimizing new software chips that can scavenge and make sense out of what they wish to glean from the natural language information clutter applying algorithms that are loosely inspired by human brain functions.

Such new tools will soon make it possible for dictators to conduct surveillance as never before, both online and in the real world.

China's Ministry of Industry and Information Technology has announced expectations that they will be able to mass-produce neural-network optimized chips by 2020. Doing so will enable China and other repressive regimes to more efficiently collect and sift through massive data on their population's speech and behavior and quickly exploit whatever information they uncover such as assessments of loyalty versus the likelihood of dissent.[244]

Deep machine learning is now training computers to identify and interpret emotional context within blocks of text using natural language processing.

Facebook now uses these techniques to examine linguistic nuances in posts that might flag users who are contemplating suicide, and smaller companies are working to score individual social media posts based upon attitude, emotion and intent.[245]

Predictim, a California-based AI startup, scoured text postings on Twitter, Facebook and Instagram to develop risk ratings babysitter applicants solely on the language in their social media postings. The company's automated assessment app gauged the individual's propensity to bully, to be disrespectful, or to use drugs.

Although public exposure of Predictims prying triggered a backlash, anyone who posts inappropriately private, improperly vulgar or naively unguarded information on the open media invites exploitation by salacious eavesdroppers.

University of Montreal professor Yoshua Bengio, a computer scientist known as one of the three "godfathers" of deep learning in AI, recently described to Bloomberg his concerns about the growing use of technology for political control:

> *This is the 1984 Big Brother Scenario...I think it's becoming more and more scary.*[246]

But we shouldn't assume that the benefits will accrue only to repressive governments.

Bengio, who heads the Montreal Institute for Learning Algorithms (Mila), has resisted offers to work for any large AI-driven technology company that doesn't protect user data. He has expressed particular concerns about protecting the creation of data trusts—non-profit entities or legal frameworks under which people own their data and allow it to be used only for certain purposes.

Recognizing that there are many ways that deep learning software can be used for good, it also presents great dangers from those who abuse its capabilities, Bengio advises:

> *Technology, as it gets more powerful, outside of other influences, just leads to more concentration of power and wealth. That is bad for democracy. That is bad for social justice, and the general well-being of most people.*

The digital information revolution can be appropriately viewed and celebrated as a great societal liberalizer. Simultaneously, these new technological advances which concentrate great power and influence in the hands of a few warrant serious rethinking regarding effective means to guard privacy and intellectual property rights of all free citizens.

Privacy safeguards must include legal and technical instruments to protect organizations, social and political groups and individuals from privacy piracy by all levels of their governments. This will require policies and procedures that carefully differentiate between uses for legitimate purposes (such as traditional law enforcement) versus to gain partisan or other special-interest advantages.

As I emphasize in *"The Weaponization of AI and the Internet"*, collaborations between astoundingly powerful Silicon Valley information technology behemoths and Chinese overlords of population surveillance and control portend dark omens which should greatly concern U.S. all.[247]

In an October 1, 2018 speech focused on security and economic issues at the Hudson Institute, Vice President Pence called on U.S. companies to reconsider business practices in China that involve turning over intellectual property or "abetting Beijing's oppression."

Pence said, "For example, Google should immediately end the development of the Dragonfly app that will strengthen Communist Party censorship and compromise the privacy of Chinese customers."

Responding to Pence's criticism, a Google spokeswoman simply described the company's work as "exploratory" and "not close to launching a search product in China." A logical follow-up question would be, "exploratory to what purpose?"

Meanwhile, Facebook is exploring what the New York Times referred to as "creepy patents" that will track "all most every aspect if its users lives".

One patent application uses information about how many times you visit another user's page, the number of people in your profile picture and the percentage of your friends of a different gender to predict whether you're romantically involved with anyone.

Another Facebook patent application characterizes your personality traits and judges your degree of extroversion, openness or emotional stability to select which news stories or adds to display.

Facebook filed a patent application that reviews your posts, messages, and credit card transactions and locations to predict when a major life event such as a birth, death or graduation is likely to occur.

Facebook will be able to identify the unique "signature" of faulty pixels or lens scratches appearing on images taken on your digital camera to figure out that you know someone who uploads pictures taken on your device, even if you weren't previously connected. They will also be able to guess the "affinity" between you and a friend based upon how frequently you use the same camera.

There is a Facebook patent application that uses your phone microphone to identify the electrical interference pattern created by your TV power cable to reveal which television shows you watched and whether you muted advertisements.

There are also a couple of patent applications that will track your daily and weekly routines to monitor and communicate deviations from regular activity patterns, your phone's location in the middle of the night to establish where you live when your phone is stationary to determine how many hours you sleep, and correlations of distance between your phone's location and your friend's phones to determine whom you socialize with most often.

Balancing enormous advantages of free and open information access with prudent privacy protections will present ever-greater challenges which will evolve in concert with ongoing fast-paced AI machine learning and data processing evolutions.

As Jack Clark, who directs policy for the research firm OpenAI, warns:

[W]e currently aren't—at a national or

international level—assessing or measuring the rate of progress of AI capabilities and the ease with which given capabilities can be modified for malicious purposes.

Clark adds that doing so:

[I]s equivalent to flying blind into a tornado— eventually, something is going to hit you.[248]

TOOLS AND TACTICS OF INFORMATION WARFARE

DEFINITIONAL RELATIONSHIPS BETWEEN cyberwarfare are highly contextually dependent; sometimes strategically linked, sometimes circumstantially exploratory, and sometimes remote or nonexistent...such as when hackers infiltrate computer systems and networks to steal data for financial enrichment or other purely nefarious purposes.

Blurred motives behind cyber espionage discoveries can sometimes make it difficult for government and corporate investigators to render definitive threat assessment and preemptive mitigation policy determinations. Whether perpetrated for military or economic gain, both exhibit common behaviors; namely sneaking into networks, looking for software flaws, inserting malware, and commandeering control systems.

An example of such motivational ambiguity was the June 2015 suspected Chinese state-backed theft of 21.5 million U.S. citizen records from the U.S. Office of Personnel Management. The

146

documents might have been stolen to seek information on special persons of strategic interest or potentially to target broader populations for disinformation campaigns.

Foreign propaganda campaigns can be viewed as a form of information warfare aimed at influencing citizen elections or legislative policies they favor. Such disinformation might use documents stolen by hackers and published—either complete or as modified to suit their purpose. Propagandists also use of social media (and broader media) to share incorrect stories.[249]

Western strategists are inclined—at least publicly—to view cyberwarfare and "hybrid information warfare" as separate entities, whereas Chinese and Russian cyber information-warfare applications are broadly perceived as more interdependent.

On the other hand, whereas U.S. and allied officials typically characterize foreign cyberattacks in terms of "assaults on and intrusion of cyber systems and critical infrastructure," their opposing counterparts are more apt to accuse Western cyber intelligence agencies of conducting "information wars" to undermine their regimes "in the name of democratic reforms."

Assessing Advanced Persistent Threats

Major information security breaches are termed Advanced Persistent Threats (APTs) when they are determined to have extracted massive amounts of information through highly organized and coordinated means and which pose substantial ongoing dangers.

APT teams focus their assaults on specific high-value targets which range from military jet designs—institutional financial records and client data—to company trade secrets and contract bidding strategies.

APTs are a particular nightmare scenario for any type of organization because most don't know they've been hacked until it's too late. Making matters much worse, even the following discovery, the pain is far from over. It can take many months to ascertain which machines inside the system have been infected can take months.

Take, for example, the case of an American company that hired a Pentagon-qualified computer security firm to clean its infected network after discovering that it was being targeted by an APT. A thermostat and printer in their building were caught a few months later sending messages to a server located in China.[250]

And first discovery and intervention may still be only the beginning. If the APT is truly persistent it may be but the first in a series to follow by another APT unit whose job it is to maintain an electronic foothold in the network. Their purpose, for example, might be to monitor internal emails to learn how the defenders are attempting to get them out.[251]

Some of these attacks are staged to manipulate information and systems rather than to extract information. Such "integrity attacks" are typically intended either to change the system users' perception or situational awareness of information they depend upon or to sabotage and subvert the system's physical operating devices and processes.

Such integrity attacks can be particularly insidious to detect and recognize because investigators generally rely on the same sabotaged computer systems to understand what is going on inside them.

Here, the most difficult challenge may be less about how to classify them, and more about rapidly distinguishing an integrity attack from a "confidentiality attack" as they happen in real-time. Both types exploit software vulnerabilities to gain entry into a system. It's what the attackers do when they get inside that matters most. The victims may not be able to tell whether the hackers have been observed to steal data, change data, add new data or deposit a malware backdoor "payload" for future exploits until after the action plays out.[252]

APTs are special among other cyberattacks in that they involve sophisticated teams of experts that combine high levels of an organization, complex methodologies and tools, and patient strategies. Each of these teams takes on a different role.

Much like a robber "casing" a bank, or a spy observing a military base, for example, a surveillance team engages in what is known as "target development," learning everything it can about the person or organization it is going after along with key vulnerabilities.[253]

Next, after the target is better understood, an "intrusion team" will then work to breach the system. Here, the initial target is frequently not the main prize but rather a pathway to get to the more highly-valued prize through trusted insiders with access privileges who unknowingly open the gates for them.

Once "in", unlike the typical criminal ethic of "grab what you can get," APT missions target very specific files. Often, in fact, they don't even open the files, suggesting that the previous reconnaissance was sufficiently thorough that there was no need to know contents. Someone had simply provided a specific list of collection requirements to be fulfilled and the disciplined intrusion team complied.

Brookings Institution security scholars P.W. Singer and Allan Friedman offer the example of APT attacks on several American think tanks in 2011 and 2012 seeking access to accounts of researchers who worked on Asian security issues along with their government leader contact information. The APT had begun by targeting employees with administrative rights and access to passwords.[254]

Singer and Friedman point out that, like other businesses, APT groups often conduct dry runs, and even "quality assurance" tests to minimize the number of antivirus programs that can detect them. Some scavenge email communications.

An "Operation Shady Rat" APT involved a counterfeit email attachment that opened implanted malware which created a backdoor communication channel to an outside web server. The server, in turn, had also been compromised with hidden instructions in the web page's code which were intended to enable the attackers to cover their tracks.

Other APTs track social network communications such as Facebook to find friends of individuals who have high access privileges inside a targeted system.

One particularly interesting example of social networking intrusion tricked senior British officers and defense officials into "friend requests" from a fake Facebook account claimed to be that of Admiral James Stavridis, the commander of NATO. Flattered to be considered a friend, it was an offer they didn't dare refuse.[255]

For added fake authenticity, another spear-phishing APT linked to Chinese intelligence and military units gathered details not only on targets' key friends and associates, but also mimicked farewells they typically used to sign off their emails (e.g., "All the best" vs. "Best regards" vs. "Keep on Trucking").[256]

Sometimes the attackers get sloppy. This occurred when a high-ranking Chinese military official was caught using the same server both to communicate with his mistress and to coordinate the APT.

APT "Exfiltration teams" enter the scene after the target has been "pwned" (at their mercy). Their job is to sweep up all the information or compromise all the systems that the APT had originally been set up to accomplish.

Not all APTs, on the other hand just copy the data and exit. Some add technology to allow them to steal new secrets beyond what was inside the network, or even gain control.

A common cover-up tactic is to rout stolen data through way stations in multiple countries. This makes the APT source identity much more difficult to track down—just as a money launderer runs running stolen funds through banks all over the world. Also, routing APT activities through different countries and legal jurisdictions can also complicate the ultimate prosecution of accused parties.

APTs that alter files, as well as steal data, can have major consequences that blur definitional boundaries between acts of crime, espionage, sabotage, or even warfare. And sometimes, they make events rising to the level of appearing like warfare by simply exploiting gross carelessness on the part of government and political hacks who get hacked for lack of sensible precautions.

Trophy Spear-Phishing of DNC and Clinton

In the middle of 2015, during the start-up of the 2016 presidential primary season, the Democratic National Committee asked former Clinton and Bush administration national security council counterterrorism chief, Richard Clarke, to assess the political organization's digital vulnerabilities.

Clarke, a hard-bitten Washington national security warrior, immediately recognized that DNC an obvious cyber espionage target was poorly prepared for defense. The basic service they had hired to filter out potential phishing spams wasn't even as sophisticated as what Google's Gmail provides; it certainly wasn't any match for a sophisticated attack. In addition, the DNC had barely trained its employees to spot a spear-phishing attack; this despite well-publicized Chinese and Russian intrusions into the Obama campaign computers in 2008 and 2012.[257]

By the time Clarke arrived at the cyber scene, it was already too late.

It was soon discovered that the DNC had been hacked by not just one Russian intelligence group, but two. And both left plenty of fingerprints.

The first was "Cozy Bear", one that the FBI associated with a group known as the "Dukes." Still, others called the group "APT 29" (for advanced persistent threat.)

In March 2016, the Brits discovered that a second competing group affiliated with the Russian GRU military intelligence unit was involved as well.[258]

An FBI agent had even contacted the DNC to inform them about the penetration. A part-time employee who took down the telephone message and emailed it to his colleagues. He wrote,

> *The FBI thinks DNC has at least one compromised computer in its network and the FBI wanted to know if the DNC is aware, and if so, what the DNC is doing about it.*

The DNC did not respond...nor even bother to return repeated FBI agent phone calls on the matter.

The agent called again in November 2015, explaining that the penetration situation was worsening. One of DNC's computers—and it wasn't clear which one—was transmitting information out of its headquarters. He warned that the machine was calling home, where "home" meant Russia.

That second warning should have set off alarms—but there is no evidence that it did, or as later claimed that the information never made it to the DNC's top leadership. Headed by Debbie Wasserman Schulz.

Topping the Russian's hit list was Clinton's campaign chairman, John Podesta, a widely-connected Washington insider who had served as Bill Clinton's chief of staff and who had organized many campaigns. A fake inbox message ostensibly from Google on March 19th 2016, had warned him that someone was trying to break into his personal Gmail account.

Podesta, who at the time was busily focused on fund-raising and message-sharpening for the Clinton campaign, had delegated a handful of his aides managed his email for him. The individual who noticed the email declaring that Podesta should change his password then passed the message along to a computer technician for a legitimacy assessment.

The technician rendered advice back to the original aid that received the phony alert that it was *not* legitimate. However, in reporting back to Podesta, a typo in the response erroneously changed the determination from "illegitimate" to "legitimate."[259]

Consequently, Podesta's password was changed immediately, allowing the Russian hackers access to sixty thousand emails stretching back a decade. That accomplishment was relatively modest, however, considering that the server was only slightly larger than a laptop computer.

Meanwhile, the communications failures between the DNC and the FBI had given the Russian hackers lots of time to review endless DNC email accounts revealing personal complaints about bosses, their worries, and their contemplation of future personnel moves. But far more interesting, they had soon moved on to pursue a far richer treasure trove outside the DNC.

Using credentials stolen along the way, the hackers stole Democratic Congressional Campaign Committee login details of a system administrator who had "unrestricted access" to the network. Doing so, they hacked into 29 computers belonging to the Campaign Committee, and more than 30 owned by the DNC.

Remembering that Chinese hackers had previously broken into both Obama and John McCain's 2008 campaign, the Clinton team had originally brought in some serious cybersecurity expertise. The result was that the campaign's networks repelled several attacks, none of them wildly sophisticated.[260]

Hillary's personal email security setup—the one that she used for a great deal of her official State Department communications—was quite a different matter. The system consisted of a series of unsecured BlackBerry phones, a jerry-rigged computer and what investigators referred to as a make-shift "homebrew" email server.

As reported in *Forbes.com* by my friend Paul Roderick Gregory, Russian intelligence knew about Clinton's private email account since—or even well before—mid-March 2013. Its existence came to light with the publication by a Kremlin-funded news service of emails sent to Clinton by her trusted advisor Sidney Blumenthal.

Thereafter, the mainstream media lost interest in the story but was revived when Clinton's private emails became the subject of congressional investigations.[261]

A subsequent FBI examination of Clinton's "extreme

carelessness" in handling official exchanges found over 100 emails containing classified information, including 65 emails deemed "Secret" and 22 deemed "Top Secret." An additional 2,093 emails not marked classified were retroactively classified by the State Department.

Over the course of five years, those emails first lived in her Chappaqua, New York basement, then later in a data center in New Jersey, then were FedExed across the country and possibly copied onto a thumb drive before being printed out, sorted and handed back to the State Department in 12 bankers' boxes.

There is a good reason to believe that Vladimir Putin might very well have held special personal animus towards Hillary which contributed to her cyber targeting motivation.

National security policy researcher David Sanger at Harvard's Kennedy School of Government points to an event in December 2012 just after Russia had just completed a parliamentary election which was riven with ballot-tampering and fraud. Putin won, but not by much.[262]

Just days after Putin's 2012 reelection, Hillary had issued a bland State Department declaration right out of the standard State playbook: "The Russian people, like people everywhere, deserve the right to have their voices heard and their votes counted."

Although not mentioning Putin or his party by name, he got the message. Clinton and her aides apparently didn't realize that calling out Russia for antidemocratic behavior was anything special. Putin, however, took the declaration very personally.

Hillary has since attached great blame for her failed 2016 presidential election bid to Russian interference. Both she and DNC chairwoman, Debbie Wasserman Schultz wasted no time using the saga to gain explanation and sympathy for having had the Russians— aided and abetted by President Trump—steal their deserved victory.

Election day came and went with no evidence of suspicious cyber activity at the polls. Even the top members of the former Obama administration appeared to have had a difficult time getting excited about the hack resisting demands from the DNC that the government does a quick public "attribution," naming the Russians as offenders.

As David Sanger recalls:

> *Even then, in the wake of Trump's astonishing win, Obama still could not bring himself to take strong action against Putin, the oligarchs, or the GRU. He worried that the United States would lose the moral high ground.*[263]

According to one former senior Obama official, all the decisions that had been made about pushing back against the Russians required reexamination:

> *There had been an assumption that Hillary would win and we'd have time to figure out a set of actions that could be carried into the next administration. Suddenly we had to come up with some steps that couldn't be reversed.*[264]

By October 2-15, Obama had concluded that the DNC/Hillary campaign attacks were more than just espionage; they constituted an attack on American values and institutions—to be viewed as attacks on free expression. Nevertheless, he privately told his aids that publicly calling the Russians out on this would appear too partisan. It might also play into Putin's hands by conceding, before a single ballot was cast, that the election had been compromised.

So, as David Sanger observed, the White House developed a two-part plan: get the leaders of Congress, Democrats, and Republicans, to issue a joint statement condemning Russia's actions, and have Obama confront Putin at a summit meeting they were both planning to attend in early September. Sanger notes:

> *The accounts of how strongly Obama threatened Putin depend on who was telling the story. But the essential warning was that the United States had the power to destroy the Russian economy by cutting off its transactions—and would use that power if American officials believed Russia interceded in the election.*[265]

Also, according to Sanger:

> *Obama emerged from the session wondering aloud whether Putin was content to live with 'a constant, low-grade conflict.' He was specifically referring to Ukraine, but he could have been talking about any of the arenas in which Putin relished his role as a great disrupter.*
>
> *It seemed clear that to Putin, low-grade conflict was just fine; it was the only affordable way to restore Russia's eminence on the global stage.*

Public attention to leaked DNC content kicked in with the first massive WikiLeaks dump of 44,000 emails and more than 17,000 document attachments. Not coincidentally, the deluge started right before the Democratic National Convention in Philadelphia.

The emails released in the trove were so blunt and insulting that they played to the divisions within the Democratic party, just as Sander's delegations were showing up in the sweltering heat of Philadelphia. The most politically potent of the emails made clear that the DNC leadership was doing whatever it could to make sure Hillary Clinton got the nomination and Bernie Sanders did not.

Ensuing Sanders' camp uproar over the DNC's Hillary bias caused Florida congresswoman Wasserman Schultz to resign as the party's chair just ahead of the convention over which she was to reside. Although cutting no evidence, Clinton's campaign manager,

Robby Mook attempted to exercise damage control by accusing the Russians of leaking the data "for the purpose of helping Donald Trump."[266]

Days later, the Obama administration leaked a preliminary assessment represented as deeply classified which concluded with "high confidence" that the Russian government was behind the theft of emails and documents from the DNC. This was the first time that the government began to signal that a larger plot was underway.

Still, due to disagreement among intelligence agencies, the White House remained officially silent on the matter. The CIA's "high confidence" was based in part on human sources inside Russia. The NSA, however, was not prepared to sign on; it did not have enough signals intelligence and intercepted conversations to say anything more than "moderate intelligence" the hack was a GRU operation, and that Putin had ordered it.[267]

Obama had secretly ordered a national Intelligence Estimate by an independent group, the National Intelligence Council," to assess the susceptibility of the American election system to outside influence.

Some holes in the system in a few of the most critical swing states were well known. Pennsylvania, for example, had almost no paper backup for its voting machines to enable a post-election audit in the event of a vote reporting challenge.

In June, it was discovered that passwords belonging to an Arizona election official had been stolen. Fearing that a hacker might use them to get inside the registration system, they took the registration database offline for ten days to conduct a forensic analysis to determine whether the data had been changed.[268]

Another forensic examination suggested that voter information had been siphoned off from the Illinois voter registration system, potentially a hack engineered by known Russian groups. The fears, while rampant, were still based on conjecture. Russian hackers had essentially been caught scouting the systems, but not changing anything.[269]

In any case, the National Intelligence Council ultimately concluded that hacking the election machines themselves on a broad scale, while not impossible, would be a daunting job. Most voting machines were offline, meaning that hackers would need a physical presence in key polling places to interfere with results.

Obama purported to lay out the facts regarding attempted Russian election interference about to reporters in mid- December 2016, stating that he didn't know about the hacking of the DNC until "the beginning of the [previous] summer." He didn't bother to mention that it was months after the FBI had made the first call to DNC headquarters.

When headline news of the DNC hacks broke out, the Obama White House response had been, according to Sanger, right out of the diplomatic playbook. Thirty-five Russian "diplomats" were thrown out of the country, most of them spies, some suspected of abetting the hacking into American infrastructure.[270]

Besides, a few Russian facilities were closed. One of these was their consulate in San Francisco, where black wisps of smoke rose from the chimney as the Russians burned paperwork.

Diplomatic properties in Long Island and Maryland were also shuttered. One of them—the Obama administration didn't say which—had been used by the Russians to bore underground and tap into a major telephone trunk line that would presumably give them access to both phone conversations and electronic messaging—and perhaps another pathway into American computer networks.

The FBI has said that its investigation was hindered by the DNC, which would not allow the FBI access to its main servers. Instead, the FBI had to rely upon secondhand evidence provided by the DNC's privately contracted Google-owned CrowdStrike investigators.[271]

In his extensive report into investigations into potential Trump campaign collusion with Russian interests, Special Counsel Robert S. Mueller noted that his team did not "obtain or examine" the DNC servers in determining whether they had been hacked by Russia.[272]

The DNC, upon refusing to provide FBI access to its servers, transferred that Russian hacking determination to CrownStrike, an arm of Google's parent company, Alphabet Inc. Eric Schmidt, CEO of Alphanet, in turn, was an active supporter of the Hillary Clinton campaign and a long-time DNC donor.

Breitbart News has reported that Perkins Coie, the law firm that represented the DNC and Clinton's campaign, helped to enlist CrowdStrike to aid with the DNC's allegedly hacked server. On behalf of the DNC and Clinton's campaign, Perkins Coie also paid the controversial Fusion GPS firm to produce the infamous, discredited anti-Trump dossier compiled by former British spy Christopher Steele.

Mueller's report stated that the GRU, the Main Directorate of the General Staff of the Armed Forces of the Russian Federation, "stole approximately 300 gigabytes of data from the DNC cloud-based account." [273]

The GRU also targeted "individuals and entities involved in the administration" of the presidential election, the report documents.

It states: "Victims included U.S. state and local entities, such as state boards of elections (SBOEs), secretaries of state, and county governments, as well as individuals who worked for those entities."[274]

Yet the Special Counsel's office *"did not investigate further"* the evidence it said it found showing that Russia's GRU targeted the DNC and the other entities. Nor did Mueller's team ever examine or obtain the DNC's servers.[275]

Had Mueller's investigation delved further, they would have found some evidence of Russian campaign tampering after all. According to DNC officials, the "hackers" had gained the entire database of opposition research on then-GOP presidential candidate Donald Trump.[276]

As for postured sanctimony regarding foreign election tampering, the United States doesn't come to such charges with clean hands either. Italy and Iran were notable targets for CIA election manipulation and coup-organizing in the 1950s. Numerous CIA attempts to kill Castro followed in the 1960s.

Putin has condemned the United States for mounting covert campaigns to influence elections in South Vietnam, Chile, Nicaragua, and Panama, along with American financial backing of pro-Western revolutions in Georgia, Kyrgyzstan, and Ukraine in the early 2000s, as well as the Arab Spring.

"Put your finger anywhere on a map of the world," Putin said in 2017, "and everywhere you will hear complaints that American officials are interfering in internal election processes."[277]

Former CIA chief of Russian operations Steven l. Hall who retired in 2015 following 30 years with the agency said: "If you ask an intelligence officer, did the Russians break the rules or do something bizarre, the answer is no, not at all."

Hall added that the United States "absolutely" has carried out such election influence operations historically, and I hope we keep doing it." [278]

American intelligence scholar Loch K. Johnson, who began his career in the 1970s investigating the CIA as a staff member of the Senate's Church Committee, agreed with Hall that Russia's 2016 operation was simply the cyber-age version of standard United States practice for decades. It was something that occurred whenever American officials were worried about a foreign vote.

Johnson said:

> We've been doing this kind of thing since the C.I.A. was created in 1947... We've used posters, pamphlets, mailers, banners—you name it. We've planted false information in foreign newspapers. We've used what the British call 'King George's cavalry': suitcases of cash.[279]

Loch Johnson was likely referring, for example, to bags of cash delivered to a Rome hotel for non-Communist Italian candidates from the late 1940s to the 1960s, scandalous stories leaked to foreign newspapers to swing an election in Nicaragua, and millions of pamphlets, posters, and stickers printed to defeat an incumbent in Serbia.

Richard M. Bissell Jr. who ran the CIA's operations in the late 1950s and early 1960s casually noted in his autobiography of "exercising control over a newspaper or broadcasting station, or of securing the desired outcome in an election."

As disclosed in *The New York Times*, a self-congratulatory declassified report on the CIA's work in Chile's 1964 election boasts of the "hard work" the agency did supplying "large sums" to its favored candidate and portraying him as a "wise, sincere and high-minded statesman" while painting his leftist opponent as a "calculating schemer."[280]

Over time, covert CIA election influence campaigns came to operate more openly by the State Department and its affiliates. The U.S. provided political consultants and millions of campaign stickers in support of a successful 2000 election win that defeated the Serbian nationalist leader Slobodan Milosevic.[281]

America was less successful to help defeat the reelection of Afghanistan President Hamid Karzai in 2009. Then-Secretary of Defense Robert Gates later referred to this failure in his memoir as "our clumsy and failed putsch."

In actuality, the 2016 Russian DNC hacking campaign primarily applied old-school international espionage that exploited the advantages of contemporary technologies. These new tools are familiar, to be expected, and to prudently be prepared for by all sides.

A key difference, perhaps, is that whereas the old-school methods stuffed suitcases with cash, modern approaches now stuff computers with malware and misinformation to achieve similar ends.

The *New York Times* national security writer David Sanger observes that the Russian DNC hacks exposed more than the Obama administration's lack of a playbook for cyber conflict. Years of Russia's ever-escalating, ever-more multifaceted attacks underscored the White House failure to anticipate that they can be used to "fray the civic threads that hold together democracy itself."[282]

Even before Putin's hackers broke into the DNC, their attention had been focused on two juicy, high-value targets; the State Department and the White house.

Their first strike was against the State Department's unclassified "low side" email system, which unlike the separate classified "high side" network, connects to the outside world. This open system enabled a convenient means for the Russians to remotely insert a malware link to their command-and-control server. Whenever State Department staffers clicked on the phishing emails, some purporting to be from American universities, the hackers were in.

With luck, the Russians may have also found clues about how to get into the "high side" classified systems. Spear phishing expeditions on high-ranking officials, including Clinton, indicated this to be the case.

A State Department team, including the FBI and NSA, eventually chased the Russians out of the system, but only to discover that they had already moved on. They soon turned up only a mile away—inside the White House servers.

The White House cyberattack turned out to be somewhat of a digital ambush for their U.S. intelligence trackers.

Their nemesis, the Russians cyber spies, were playing a game of identity hide and seek from command-and-control servers they had placed around the world.

As David Sanger observed,

> *Every time the NSA's teams of hackers cut the links, they found that the White House computers began communicating with new servers. No one had seen anything quite like it—a state-sponsored*

group of hackers in a digital dogfight.[283]

To the NSA's team, it looked a bit like a video game with real-life consequences, and they appeared to be having a good time in gaming the NSA.

Sanger notes that the battle for control of the computer networks at the State Department and the White House raised two big questions.

First, he asks, "why did the Russians choose to take on the United States so directly?"

And second, "why did the Obama administration try to keep the whole series of incidents secret, including the hacks on the Pentagon and Congress?"

Sanger postulates that the answer to the first question seems simple: namely that they wanted the U.S. to know that they were in the big leagues. In short, the Russian hackers were strutting their stuff for the same reason that Russian generals parade their tanks and missiles just across the border from Lithuania.

The bigger mystery, Sanger ponders, is why Obama once again chose not to call the Russians out, referring to the incidents at a White House Situation Room meeting with intelligence officials merely as "just espionage."

Sanger concludes:

> And if the United States was sloppy enough to let it
> happen, then the answer was to up our game on
> defense rather than think about retaliation.

Inside Putin's Propaganda Provocateurs

US and allied understanding of Russia's espionage games and players has benefitted from some quiet help from the Netherlands.

According to an investigation by two Dutch news organizations, the tiny nation's intelligence agencies had penetrated a university building off Red Square in Moscow where the Cozy Bear group operated.

The Dutch had not only gotten into Cozy Bear's computer systems, but also into their building's security cameras. This enabled the Dutch intelligence service to see what the Russians were doing and, assisted by facial recognition software, to identify the operators who were doing it.

Still, Russia's Internet Research Agency (Glavset) which is believed to have initiated the DNC, Clinton and other foreign cyber espionage and political tampering attacks remains to be a recent and highly mysterious government-sponsored enterprise.

David Sanger reports, the now multi-billion annual budget organization first got off to a fast start in Saint Petersburg and began hiring some time in 2013. Its hires included a variety of backgrounds and skills, including news writers, graphics editors, and experts in "search-engine-optimization."

To ensure the greatest outreach for its pro-Russian messages, Glavset took advantage of the fact that Facebook, a popular platform tool, did little (at least at the time) to ascertain whether a member was a person—or just a bot.[284]

In late 2014, the agency launched a social media campaign to generally disrupt upcoming U.S. elections. Early "beta" efforts involved sending hundreds of fake information messages on Facebook and thousands on Twitter targeting prospective voters on both sides of such contentious issues as immigration, gun control, and minority rights.

Texas seemed particularly ripe for meddling. Although few of the trolls and bot makers had likely been there ever, they must have heard much about it online and formed images about its political culture from movie depictions.

A fictitious "Heart of Texas" online group operating out of facilities near Red Square also established an opposing group called "United Muslims in America." Then, in a masterstroke, the schemers arranged and scheduled a counter-rally of the second against the former fake organizations under the banner of "Save Islamic Knowledge."

The waggish idea was to motivate actual Americans—who had joined each of the Facebook groups—to face off against each other and incite a lot of name-callings and, hopefully, some violence as well.[285]

Glavset operators then moved on to advertising. Between June 2015 and August 2017, the agency and groups linked to it spent thousands of dollars on Facebook propaganda ads each month—at a fraction of the cost for an evening of television advertising on a local American television station.

In doing so, Putin's trolls reportedly reached up to 126 million Facebook users and made as many as 288 million Twitter impressions. It is unknowable if or how the campaign had any significant voter impact.

Meanwhile, U.S. national security investigations into the purposes and players of responsible for hacking the DNC became even more intriguing with the arrival of a persona with the screen name of "Guccifer 2.0" burst onto the web scene claiming that he— not some Russian group—was responsible. Instead, he contended that he was simply a very talented individual hacker.

The cyber culprit's chosen moniker was taken from "Guccifer, the alias of a Russian hacker who was then sitting in jail, after famously breaking into the email accounts of former Secretary of State Colin Powell and former President George W. Bush.

David Sanger maintains that anyone who had followed the Russian hacking groups knew immediately that there was little chance that Guccifer 2.0 was simply a savvy, lone hacker. As for his regional origin, however, his awkward English, which became a hallmark of the Russian effort, also made it evident that he was not a native speaker. He wrote:

> *Worldwide known cybersecurity company CrowdStrike announced that the Democratic*

National Committee (DNC) servers had been
hacked by 'sophisticated' hacker groups.

I'm very pleased the company appreciated my
skills so highly. But in fact, it was easy, very easy.

Guccifer may have been the first one who
penetrated Hillary Clinton's and other Democrats'
mail servers. But he certainly wasn't the last. No
wonder any other hacker could easily get access to
the DNC's servers.

Shame on CrowdStrike: Do you think I've been
in the DNC's networks for almost a year and saved
only 2 documents? Do you really believe it?[286]

For credence, Guccifer 2.0 offered a sampling of purloined
documents, which included a lengthy piece of opposition research
prepared by the DNC with chapter headings like: *"Trump Is Loyal*
Only to Himself" and *"Trump Has Repeatedly Proven to Be Clueless*
on Key Foreign Policy Issues." There was also a chart listing major
DNC donor, where they lived, and how much they had contributed.

Guccifer 2.0 then warned that the samples were just a tiny
part of all documents he had downloaded from the Democrats'
networks, adding that "thousands of files and mails," were now in
the hands of WikiLeaks.

"They will publish them soon," he predicted.[287]

Online cyber sleuths soon determined that Guccifer 2.0 was far
more likely to be a committee of hackers linked to the Russian GRU
than a single independent agent.

Vice magazine investigator/writer Lorenzo Franceschi-
Bicchierai came up with the bright idea of sending Guccifer 2.0 a
direct message. Doing so, Guccifer 2.0 immediately responded,
claiming to be Romanian.

So Franceschi-Bicchierai then used Google Translate to ask
Guccifer 2.0 some questions in stilted Romanian. When the answers
came back in equally stilted Romanian it became apparent that

Guccifer 2.0 was using Google Translate too. Whoever he was, he didn't speak the language either.[288]

A closer look at the documents posted by Guccifer 2.0 revealed that they had originally been written in a Russian version of Microsoft Word, and then edited by someone who perhaps cynically identified himself as Felix Edmundovich. Interestingly, a fellow named Felix Edmundovich. Dzezhinsky was the famous founder of the Soviet secret police. In fact, Dzerzhinsky Square in Moscow, where the KGB headquarters was located, was named after him following the fall of the Soviet Union.

Growing evidence led Franceschi-Bicchierai to become more and more convinced that he was dealing with a group of people who were either not very good at covering their tracks, or that didn't seem to wish to cover them in the first place.

In the meantime, more hacked documents suddenly appeared in *"DC Leaks,"* an online site established just a few months before, but not previously active. This second outlet offered another indication that making selected stolen documents public was part of a larger plan that had been formulated months in advance.

Even more strategically suspicious was the fact that the hackers—presumably Russian—had taken plenty of time reviewing the stolen Podesta emails obtained in March 2016 before releasing them and exposing the penetration. This interlude enabled numerous very patient and dedicated professionals to carefully sort through enormous numbers of documents to determine which were most valuable and scandal newsworthy. Included were undisclosed texts of Hillary Clinton's speeches to Goldman Sachs which she had refused to publicly release.

Other investigating organizations and agencies also found even more suspicious GRU links to the DNC hack.

In the spring of 2016 Britain's GCHQ, their equivalent to the NSA intercepted an interesting series of messages which appeared to have originated from DNC computer servers plucked out of Russian metadata networks. That information trove was clearly traced to the

GRU, whose activities GCHQ constantly monitored.

Following the discovery, the GCHQ notified the NSA where a senior-level official responded that they appreciated the heads-up. That was reportedly the last response GCHQ received from NSA on the topic.[289]

The *New York Times* investigator David Sanger concludes that, for obvious reasons, no one will be very precise about how the British picked up the traffic that led back to the DNC. Nevertheless, there are several clues.

Sanger publicly reported that Snowden documents reveal that the GCHQ was plugged into two hundred fiber optic cables and that they possessed remarkable capabilities to process mostly-encrypted metadata from forty-six of them simultaneously.

Significantly, Britain is a critical hub—termination point—for cables that cross into Russia come ashore. Intelligence agencies, both the United States and Britain, pay "intercept partners"—like AT&T and British Telecom—to keep teams of technicians at termination sites to mine and hand over data.[290]

According to Sanger, there are similar listening posts around the world, which are divided up for monitoring among "Five Eyes" countries: Australia, Canada, New Zealand, the United Kingdom, and the United States:

> *While the Brits focus on Europe, the Middle East, and western Russia, the Australians monitor East Asia and South Asia—which is why operations in Afghanistan are often run out of Pine Gap, in the Australian desert.*
>
> *New Zealand owns digital traffic in the South Pacific and Southeast Asia. The United States, with huge collection budgets, looks at hot spots, starting with China, Russia, Africa, and parts of the Middle East.*[291]

Sanger explains that such monitoring is quite naturally a subject that officials in each of these countries won't discuss openly—not even years after the Snowden WikiLeaks revealed them to the world. One reason for this reticence is because that these termination points are no longer just places to plug in headphones. They also provide locations and means to inject implants—malware—into foreign networks.

Monitoring and protection of metadata termination points are vitally critical to ensure national security from all foreign disruptors. Destructions or seizures of six or so of these locations could slow information flow to, from and within the United States to a trickle. This means, for example, that virtually all telephone and news communications could be compromised or disabled. Such protection imperatives have prompted Facebook and Google to begin laying their own cables.

NSA Bugged by its Own Inventions

In mid-August of 2016, a time when Democrats were still struggling to figure out what the Russian hackers were doing to them, a group calling themselves "Shadow Brokers" posted cyberattack tools, including malware codes and instruction manuals, designed to exploit Microsoft system vulnerabilities.

The NSA immediately recognized the malware code. After all, they had written it as a means to place implants in foreign systems where it could lurk undetected for years—at least it could hide so long as the target didn't know what to look for.

Now foreign cyber sleuths everywhere could not only spot the temporarily dormant bugs. Shadow Brokers had offered the world a virtual product catalog.

David Sanger reports that within the NSA, the real security damage caused by Shadow Brokers' secrecy breach eclipsed even that of the more publicly well-known Snowden affair. Whereas Snowden had released code words and PowerPoints describing what amounted to battle plans, the Shadow Brokers had posted the actual code cyber weapons.

And the Shadow Brokers releases just kept coming; highly classified technical information wrapped in taunts, broken English, a good deal of profanity, and numerous references to the chaos of American politics. They promised a "monthly dump" of stolen tools, leaving transparent clues—perhaps intentional misdirection—that Russian hackers were behind it all.

One such missive read: Russian security peoples, is becoming Russian hackers at nights, but only full moons.[292]

Explanations regarding how the secrets most likely ended up in Russian hands are linked to two incidents, both involving NSA-contractors.

When first occurred in late 2014 or early 2015, beginning in about 2010, a Booz Allen Hamilton NSA contractor employee working deep inside the agency began to bring home classified documents, many in digital form.

There is no evidence that the individual, Nghia H. Pho, did so with any covert intent. Nevertheless, the documents he removed had apparently been contaminated with infected antivirus software triggers designed to instruct any computer it was installed on to search for NSA code words and communicate the information through a back door to Russian intelligence.[293]

Then in early October 2016, FBI investigators trying to crack the Shadow Brokers case arrested another Booz Allen Hamilton contractor, Harold Martin III, whose house and a car in the suburban tract in Glen Burnie, Maryland were brimming over with classified documents, including many terabytes of electronically stored information.

Court filings revealed that although many of those materials were from NSA, others had been stolen from the CIA, Cyber Command, and the Pentagon.

Whereas Martin was not ultimately found to have worked for the Russians, the materials found in his possession did include some tools that were ultimately put up for sale by Shadow Brokers. This troubling discovery left open concerns regarding further leaks from NSA systems.[294]

DARK WEBS OF TRAPS AND TERRORS

DANGEROUS CYBER CULPRITS lurk in the dark depths of the "World Wide Web" which can be likened to an ocean, a vast territory of unknown and inaccessible locations to the average user. An estimated 95% of that ocean is "invisible" to typical users on its "surface" who rely upon traditional search engines, yet it comprises only about 3% of overall Internet traffic.[295]

The "visible web", on the other hand, is readily accessible to the general public surface surfers. It's the more familiar part that consists of sites whose domain names end in .com, .org, .net, or similar variations.

Stated another way, the World Wide Web is an information-sharing model built on top of the Internet that uses the HyperText Transfer Protocol (HTTP) which defines how messages are formatted and transmitted plus what actions web servers and browsers such as Chrome or Firefox should take in response to various commands to share information.

Accordingly, the web is a large part of the Internet, but not it's only component. For example, email and instant messaging are not part of the web, but rather, are part of the internet.[296]

The "invisible" or "hidden" web refers to all digital content that cannot be found with a search engine: information including a Gmail account, online bank statements, office intranets, direct messages through Twitter, and photos uploaded to Facebook marked "private."

The invisible web isn't hidden for any inherently bad reasons. Governments, researchers, corporations, personal and political blogs, news sites, discussion forums, religious sites, and radio stations, for example, go there to store content on dynamic web pages.

People regularly use deep content without realizing it…material that is dynamically produced via accessing a surface site as a unique web page seen only upon request. Examples here include travel sites such as Hotwire and Expedia which allow searchers to directly access airline and hotel databases through a query in the search box, such as the name of a destination.

The web ocean gets darker as we dive deeper. Much darker.

The layer just below the surface web lies what is sometimes referred to as the "deep web", a private area that while isn't accessible by typical search engines, simply contains that protected password-protected information that isn't available to the public. These data components include personal bank accounts, retail accounts, and member-only sites and internal school or company websites.

Then, diving much deeper, we reach "dark web" depths which can only be accessed with specific software. This region which comprises about 3% of the Internet is a particularly popular hidden domain for criminals to lurk: a place where very stealthy thieves who remain anonymous and untraceable collect and stash private information stolen from the level above.

Dangerous Predators at Dark Depths

The dark web is a hidden network of websites which, unlike user IP addresses which can be traced to a computer, enable visitors to mask their true identities. This region can only be accessed through networks such as Tor ("The Onion Routing" project) which provides anonymous access to the Internet, and I2P ("Invisible Internet Project") which specializes in allowing anonymous hosting of websites.

User anonymity is accomplished by masking software that takes an evasive randomized path that bounces between several encrypted connections to its ultimate file destination. This complicated strategy makes it nearly impossible to trace a node path between a large number of intermediate servers and decrypt the information layer by layer.

Also referred to as the "darknet", this region is a spawning grounds and truly black marketplace for DDoS botnet recipes, hacking group services, phishing scam gamebooks, bitcoin money laundering capers, child pornography, and sex traffickers, illegal drug connections, sales of fake IDs and passports, and recruitment and coordination of terrorist enterprises.

A 2016 report by the European intelligence organization Intelliagg and its U.S. counterpart, DARKSUM, estimates that about half of the dark web content is legal under U.S. or UK law.[297]

Global networks of dark web fraudsters gain access to valuable information such as Social Security Numbers and bank account access codes captured through data breaches and hacks. This information is often sold to third-party purchasers in bundles called "Fullz" (full packages of individual private materials which may include SSNs, birth dates, account numbers and other data that can be hijacked.)

Fullz is frequently traded on the dark web repeatedly. Entire "communities" of traders even post reviews that indicate whether or not the criminal source is someone "good to do business with."

The dark web can be a very dangerous trap for unaware Internet users who allow themselves to be pulled in. According to Michael Lewis, a blog writer with Moneycrashers,"Browsing its hidden sites without precautions might be compared to trying to get safely through a village infected by Ebola." [298]

In addition to malware dangers, dark web visitors to controversially dangerous sites may invite unwanted government surveillance. This applies particularly to countries such as Iran, Syria, and Turkey were browsing a government-monitored ideologically out-of-line website can get you into really big trouble.

An unceasing technology race is pitting the "good guys" who are developing tools to reveal unlawful activities against those who are producing ever more sophisticated new generations of malware. We can thank some of the Big Tech companies for leading contributions to this ongoing defensive warfare.

At the same time, even many proponents of Internet privacy assert that the dark web is essential to protect our freedom and liberty. In support of this view, a 1997 U.S. Supreme Court decision in the case of *Reno v. American Civil Liberties Union* extended full First Amendment protections to the Internet virtually guarantees its continued existence.

Global Bonnies and CyberClydes

"Fancy Bear", the Russian cyber group tied to the famously stealing and leaking DNC emails, wasn't new at the game of trying to influence foreign national elections. Their efforts to manipulate democracies have also been discovered in German, French, and Ukrainian campaigns. [299]

Operating under a variety of different identities, Fancy Bear reportedly got its start in 2008, hacking the Georgian government to throw it into chaos just before the Russian government invaded the country. Since then, they've been involved in countless controversies and conflicts in the region, doing everything from threatening anti-

Kremlin journalists and protesters, hacking the German parliament for over six months in 2014, making death threats to wives of U.S. Army personnel, disabled 20% of Ukraine's artillery via a corrupted app.[300]

Russian cybercriminal Evgeniy Mikhailovich Bogachev certainly appears to be connected with the Russian government as well. It took the FBI and other international crime organizations two years just to get the guy's name after he had managed to infect millions of computers around the world with ransomware, steal all of their stored data and cause an estimated $100 million in damages.

Although the FBI offered three million to anyone who can help bring Bogachev to justice, the Russian government has shown no interest in collecting. Although Moscow has never admitted working with him, the cybercriminal lives openly in Anapa, a run-down resort town on the Black Sea in southern Russia with a number of luxury cars and his private yacht.

In addition to establishing a form of Big Data "surveillance capitalism"—in which economic value is created and controlled through the harvesting of data about every aspect of people's daily activities—the likes of which the world has never seen. This ambitious enterprise includes huge foreign espionage and hacking operation conducted by China's Unit 61398.[301]

Between 2006 and 2011, Unit 61398's operation Shady RAT infiltrated and stole data from over foreign 70 companies, governments, and non-profit organizations.

Google was hacked in 2009, in part to spy on accounts of Chinese human rights activists and other dissidents. Five years later, Beijing stole sensitive information on around 22 million people, stored at the U.S. Office of Personnel Management, who had or previously held security clearances with the U.S. government.[302]

A hacker group believed to be backed by China has targeted universities engaged in naval research to support the country's naval capability developments. The University of Hawaii, the University of Washington, and MIT are among the at least 27 institutions hit in the US, Canada, and Southeast Asia.[303]

China also targets foreign engineering, transportation and defense sectors. Two of the group's techniques include web shells and spear-phishing emails. A web shell is a script that can be uploaded onto a web server to gain remote administrative control over it. Phishing emails usually involves embedded emails with attachments containing malware, as well as malicious Google Drive links.

Unit 61398 has been blamed for a hack that saw sensitive documents stolen regarding Israel's missile defense system. The Unit is large, estimated to have well over a thousand servers and a massive army of online hackers.

Hacking Israel's military information is a particularly big deal considering that there is likely no more sophisticated cyber intelligence organization in the world than Israel's Unit 8200.

According to Peter Roberts, a senior research fellow at Britain's Royal United Services Institute: "Unit 8200 is probably the foremost technical intelligence agency in the world and stands on a par with the NSA [the U.S. National Security Agency] in everything except scale."304

Founded in 1953 as the 2nd Intelligence Service Unit, the organization has since expanded into the largest Unit in the Israeli Defense Force. While many of their activities are clandestine, a few of their exploits have slipped to the surface. For example, Unit 8200 helped to develop the Stuxnet virus aimed at Iran's nuclear plants, along with very advanced spying malware called Duqu 2.0.[305]

North Korea, a country ravaged by economic sanctions. has established a substantial infrastructure of sophisticated cyberhacking cells that primarily target corporate enterprises for financial gain. Their Bureau 121 created the WannaCry ransomware which infected around 300,000 devices in 2017 that paralyzed more than 150 organizations.

Bureau 121's most frequent U.S. marks were targeted on Houston, an oil and gas hub, and New York, a finance hub. Foreign attacks primarily targeted banks in Germany, Turkey, and the U.K.[306]

Also, recall that North Korea used hackers to destroy Sony Pictures Entertainment computer servers in retaliation for a "dear leader" comedy movie that mocked Kim Jong Un. The attack paralyzed the studio's operations, leaked embarrassing executive emails.[307]

While Iran is unlikely to match the cyber capabilities of Russia, China, or even North Korea in the short term, threats posed by this third-tier actor can't be ignored.

Between 2011 and 2013, in some of their first forays into cyberwarfare, Iranian hackers cost U.S. financial institutions many millions of dollars and knocked Saudi Aramco's business operations offline for months. These operations hit more than 40 American banks, including JPMorgan, Chase, and Bank of America.[308]

The March 2018 "SamSam Ransomware" cyberattack that struck the City of Atlanta and cost their taxpayers as much as $17 million was traced to two Iranian men. FBI investigators have charged that Atlanta was just one of several regions targeted by those criminals which hit more than 200 victims in locations that include Newark, New Jersey, the Port of San Diego, and multiple medical centers.[309]

The heist gained the two criminals more than $6 million in ransom payments with losses to victims being over $30 million.

As then-Deputy Attorney General Rod Rosenstein reported:

> *After gaining access to computers, they remotely installed their ransomware. The ransomware encrypted the computer data, crippling the ability of the victims to operate their businesses and provide critical services to their customers.*[310]

According to the *Atlanta Journal-Constitution*, the City of Atlanta had received years of warning about its cybersecurity vulnerabilities.

A 2010 independent audit account had warned that the Information Technology Department "currently does not have funding for business continuity and disaster recovery plans."311

The Atlanta Journal-Constitution also observed:

> *The city's technology leadership appears to have been overwhelmingly focused on making Atlanta a so-called 'Smart City'—a designation for cities that emphasize information and communication technology to enhance public services such as utilities and transportation.*[312]

The newspaper went on to conclude:

> *However, several cybersecurity trade publications have highlighted how these cities are especially vulnerable to attacks because of massive interdependent computer systems that constantly communicate with each other and often aren't tested before being deployed.*[313]

Mischief and Mayhem by Independents

An eternal national cybersecurity challenge is to sort out and identify real threats and intentions from suspicious communication glitches that do not. A detected stream of unusual traffic that hits a critical intelligence or infrastructure system firewall might simply be a misconfigured application or prank from somewhere in the world, or it could also be a hostile probe of defenses.

Malware packets, after all, are not like ICBMs, where radar can quickly identify the missile for what it is. Their primary design and use are guided by stealthy strategies to avoid detection and source attribution.

The dark web provides a hugely infected flea market of

malware and instruction books for government-sponsored and independent culprits who contribute to an ever-growing number of secrecy penetrations.

One of the biggest misconceptions about such hackativists is that they are all bona fide hackers with a real understanding of their actions. Actually, the vast majority are what are known as "script kiddies" who use "scripts," or programs made by others. They simply download the attack software from a website and join in with a click of a button, no expertise required.[314]

Security investigator David Singer at Harvard's Kennedy School of Government observes that a typical script kiddy uses existing and frequently well-known and easy-to-find techniques and programs or scripts to search for and exploit weaknesses in other computers on the Internet—often randomly and with little regard or even understanding of potentially harmful consequences.[315]

"Thanks" to releases of U.S. national security secrets by Edward Snowden and others, some of this information is a hugely discounted buyer's bargain. The multi-billion-dollar NSA program developed to break into computer networks from Tehran to Beijing to Pyongyang was priced on the dark market for under $100 as a piece of commercial software called a "web crawler".

Web crawlers are essentially digital Roombas that move systematically through a computer network the way that Roombas bounce from the kitchen to the den to the bedrooms, vacuuming up whatever lies in their path. They are designed to automatically navigate among websites, following links embedded in each document. If desired, web crawlers can be programmed to copy everything they encounter.

Snowden claimed that his goal in revealing inner NSA secrets was to expose what he viewed as a massive wrongdoing and overreach involving covert monitoring of Americans on U.S. soil, not just foreigners, in the name of tracking down terrorists. In doing so, PowerPoint after PowerPoint documented how NSA's Tailored Access Operations unit—known as TAO—found ways to break into even the most walled-off, well-secured computer systems around the world.

TAO workers are constantly busy designing new malware implants that can lurk in a network for months, even years, secretly sending files back to the NSA.

Sanger notes that while officially the TAO no longer exists, it has been informally absorbed into other offensive units. And although the hacking unit can't compete with Silicon Valley salaries, it continues to attract many of the agency's young stars who thrill to the idea of conducting network burglaries as a form of patriotic covert action.[316]

In addition to the centrally-orchestrated cyber infiltrations such as Snowden and WikiLeaks, other independent rogue groups conduct organized cyber attacks for independent motives ranging from political protest to vigilantism to sheer amusement.

With no single leader or central authority, such groups may visualize themselves not as a democracy or a bureaucracy, but rather, as a "do-ocracy," where members communicate via various forums and Internet Relay Chat (IRC) networks. Twitter, Facebook, and YouTube offer popular media platforms for internal debate, target identification, coordinated action planning, and global volunteer recruiting.[317]

Some attacks referred to as "doxing" (releasing documents), focus upon unearthing and publicly releasing sensitive personal or corporate information that is embarrassing to their victims. This often requires minimal network penetration, relying more on careful research to "get the dirt" on their targets.

One of the first mentions came in 2007 when Canadian news reported the arrest of a fifty-three-year-old child predator who had been tracked down and turned into the police by a "self-described Internet vigilante group called Anonymous."

In 2011, Aaron Barr, the CEO of the computer firm HBGary Federal, experienced what *Wired* magazine described as walking into "a world of hurt" inflicted by Anonymous. That pain occurred soon after Barr announced that his firm had penetrated the organization and that its investigatory findings would be released to the media at a major upcoming San Francisco conference.

The Anonymous group quickly posted a cynical response on HG Gary Federal's own compromised website. It warned:

> *Your recent claims of 'infiltrating' Anonymous amuse us, and so do your attempts at using Anonymous as a means to get attention to yourself...What you failed to realize is that, just because you have the title and general appearance of a 'security' company, you're nothing compared to Anonymous. You have little to no security knowledge...You're a pathetic gathering of media-whoring money-grabbing sycophants who want to reel in business for your equally pathetic company. Let U.S. teach you a lesson you'll never forget: don't mess with Anonymous.*[318]

The reprisal didn't end there.

In addition to taking over the firm's website, Anonymous seized control of HB Gary's email system and dumped more than 68,000 private messages and memos onto the public Internet. Embarrassingly dirty laundry included Barr's discussion of logging onto teen chat rooms and posing as a sexy sixteen-year-old girl named "Naughty Vicky."

Anonymous conducted a separate doxing attack that took control of Barr's personal Twitter account, and then used it to publicly post his Social Security number and home address.

Barr resigned in disgrace within a month when the releases prompted a congressional investigation into government contracts with the firm.

In what *Wired* described as an electric version of a final "beat down," Anonymous posted an ominous concluding message on HB

Gary's website:

> *It would seem the security experts are not expertly*
> *secured. We are Anonymous. We are Legion. We*
> *do not forgive. WE do not forget. Expect us.*[319]

Terrorists Travel Info Super Highway

The FBI defines cyberterrorism as a "premeditated, politically motivated attack against information, computer systems, computer programs, and data which results in violence against noncombatant targets by sub-national groups or clandestine agents."

The worldwide web superhighway has provided such terrorists with easy access to ultra-low cost cyberspace assets it could previously only have dreamed of a generation ago. For example, they can now scope out prospective targets using satellite-enabled Google Earth capabilities, precision target them with global positioning systems (GPS), and coordinate them with smartphones just as Lashkar-e-Taiba, a Pakistan-based terror group, did when planning its 2008 Mumbai attacks.

And just as the Internet dark web within the Internet affords a cheap global rummage sale for malware and teaching manuals, it has also provided new means for terrorist organizations to distribute a particular type of knowledge referred to by experts as "TIPPS", short for tactics, techniques, and procedures. Included are detailed designs for improvised explosive devices used across conflict zones reaching from Iraq to Afghanistan.[320]

Thus far, state-sponsored terrorist groups haven't so much attacked the Internet or used it to attack physical systems as they have used it to plan and coordinate attacks on embassies, railroads, and hotels. They have also used the Internet to raise funds, recruit and radicalize participants, and train globally-dispersed cells.

Al-Qaeda and other terrorist groups have harnessed the power of the Internet to reach the wider world in a way never before possible. This enabled Bin Laden's speeches and musings crudely recorded a hillside hideout to be seen by millions...not only by public media, audiences, but also in social chat rooms where individuals could be recruited for coordinated and "lone wolf" attacks.

Remotely distributed training over the web is reducing opportunities for international law enforcement agents to detect whereabouts of known would-be terrorist groups who no longer congregate in one place long enough for effective deterrence...a targeted cruise missile strike for example.

Cybersecurity and terrorism experts Richard Clarke and Robert Knake note that after losing their physical training grounds, al Qaeda shifted to remote virtual training camps on the web which featured such instructional videos as to how to build explosive devices out of commonly available materials and technique tips for staging beheadings.[321]

Meanwhile, rather than depending upon the dark web, much of such pro-terrorist propaganda is now open to a clear public view.

The Taliban, for example, ran a website for over a year that kept a running tally of suicide bombings and other attacks against U.S. troops in Afghanistan. And yet the host for the website was a Texas company called *ThePlanet* with 16 million accounts that rented out websites for $70 a month, payable by credit card.

Yet the company was entirely unaware aware of the Taliban site's existence and took it down once notified by authorities.[322]

As in other areas of cybersecurity, individuals and organizations must be aware of ways that their Internet habits and uses can be exploited by bad actors.

Peter Singer and Allan Friedman recount an incident in 2007 when U.S. soldiers took smartphone photos of a group of new U.S. Army helicopters parked at a base in Iraq and then uploaded them to the Internet.

Although neither the existence of the helicopters weren't deemed to be classified, the photographs revealed other information of special interest to the enemy. "Geotags" on the Internet uploads revealed where the photographers had been standing, enabling insurgents to pinpoint and destroy four of the helicopters in a mortar attack.[323]

Casually-shared Internet imagery also provides means for bad actors to identify and target individuals.

Following the 2011 bin Laden raid an American cybersecurity expert was able to determine all names and home addresses of the supposedly super-secret 12-member unit that carried it out using a series of simple social networking tricks.

One member of the raid team was identified from a website photo of his SEAL training class. Another was tagged after the curious investigator located an online image of a person wearing a SEAL team T-shirt with a group of friends, then tracked down each of them.

The cybersecurity sleuth uncovered names of FBI undercover agents using the same tactics, and in another case, discovered two married senior U.S. government officials who were participating in a swinger site (and thus vulnerable to blackmail).[324]

Fortunately, just as networking benefits of cyberspace allows terrorists to link as never before, it also allows intelligence analysts to map out social networks in unprecedented ways, it also provides clues about the leadership and structure of the perpetrators.

This is what happened to Yaman Mukhadab and to the Global Islamic Media Front (GIMF), a network for producing and distributing radical propaganda online. In 2011, it had to warn its members that the group's encryption program, *"Mujahideen Secrets 2.0,"* actually shouldn't be downloaded because it had been compromised to spy on them.[325]

In 2010, British counter-cyberterrorism operators hacked into communication networks of an al-Qaeda group in the Arabian Peninsula that was in the process of posting an English-language online recruiting and terrorism tutorial magazine called "Inspire." The Brits reportedly replaced "how to" bomb-making instructions planned for the first issue with a cupcake recipe.

In an even more lethal intervention, the instructions were changed so that the attacker would instead blow himself up during the construction of the device.[326]

Terrorism dangers posed by conventional and improvised explosive devices have taken on explosive new dimensions of reach and carnage with the advent of inexpensive, commercially available drone delivery systems.

As reported by a 2018 Rand Corporation study:

> [T]he use of weaponized drones by lone individuals and small groups—some acting as proxies of nation-states—is no longer just a concern for the future, but very much for the present. The proliferation of certain emerging technologies has effectively diffused power and made it available at the lowest levels.[327]

Very low costs of entry enabled by the proliferation of off-the-shelf hobby shop-level technology put easily acquired weaponized drone capabilities in the hands of rogue individuals as well as state-sponsored groups. This foreboding reality was ushered in with an attempted assassination of Venezuelan President Nicolas Maduro using two drones, each equipped with a kilogram of powerful plastic explosives, at a public outdoor event in Caracas.

State sponsorship of terrorist groups greatly increases the likelihood of larger and more frequent drone attacks by providing more advanced equipment and training. Hezbollah and Houthi rebels have allegedly used drones to ram Saudi air defenses in Yemen. The Islamic State has used them both to conduct surveillance and reconnaissance and for offensive actions such as dropping grenades on an adversary's military base.

The Rand Corporation warns that as more countries develop and use armed drones in combat—Nigeria, Pakistan, and Turkey have already done so—the chances for sophisticated drone technology getting into the wrong hands increased exponentially. Terrorist groups can be expected to purchase or steal drones from rogue states or corrupt military or intelligence officials along with instructions for using them.[328]

The use of drones is to deliver explosives but one concern. Law enforcement and security services universally worry about ways to defend against uses of small drones to deliver and disperse deadly chemical and biological agents, such as, for example, over a sports stadium or other public gathering place.

Although terrorist organizations such as Al Qaeda have thus far not been capable of staging a cyberattack, this circumstance may change as cost barriers to entry continue to drop. Add to this that with the advent of machine learning and artificial intelligence, drones may soon become programmable and smart enough to be used for increasingly nefarious ends with little or no human training and guidance.

We can be certain that highly adaptive terrorists will continue to find innovative new ways to spread fear and chaos. Accordingly, the Rand Corporation emphasizes an imperative that counterterrorism specialists begin planning a robust response to the threat, not only in terms of detection and countermeasure technology but also the training necessary to defend against attacks by weaponized drones.

Over the long run, laws, and policies governing drone development, use and defensive safeguards need to be established before an attack takes place, not in the tragic aftermath.

CYBERSECURITY VULNERABILITIES GO VIRAL

WHETHER INSTIGATED AND executed by rogue terrorist organizations, or by large, carefully organized, well-funded, and sophisticated state sponsors, cyberattack threats have become ever-present, escalating and alarming global reality.

Worries over vulnerabilities in critical infrastructure to cyberattack have real validity. From 2011 to 2013, probes and intrusions into the computer networks of critical infrastructure in the U.S. went up by 1700 percent.[329]

In 2001, Al Qaeda laptop computers seized in Afghanistan showed models of a dam and engineering software that simulated catastrophic failure if controls.

Logs that were discovered also revealed that Al Qaeda members visited websites that offer software and programming instructions for the digital switches that run water, power and communications facilities. During interrogations at Camp X-Ray, Cuba, prisoners told CIA officials that there had been nascent plans to use these switches to attack America.

Similarly, in 2006, terrorist websites promoted cyberattacks against the U.S. financial industry in retaliation for abuses at Guantanamo Bay.[330]

In 2011, a water provider in California hired a team of computer hackers to probe vulnerabilities of its computer networks. It didn't take very long to break in. The simulated attackers got into their system in less than a week.[331]

In the opening months of 2018, the federal government warned utilities that Russian hackers had already inserted malware implants in America's nuclear plants and power grids. Fortunately—at least thus far—understanding the control software for an electric grid, however, is not a widely available skill. It is one thing to find a way to hack into a network, and quite another to know what to do once you're inside.

Nevertheless, moderately skilled hackers are rapidly becoming more adept at causing major damage, with lots of help from codes and instruction manuals they can access from online vendors and cohorts.

Russian culprits are also believed to have been discovered infesting the routers that control the networks of small enterprises and even individual homes. There is similar evidence that Iranian hackers inside financial institutions and Chinese hackers siphoning off millions of files detailing the most intimate details of the lives of Americans seeking security clearances.

America, one of the world's stealthiest, most skillful cyber powers, is of course, applying those capabilities against our adversaries as well. Edward Snowden's leaked documents showed that the NSA had hacked the prestigious Tsinghua University in Beijing—home to one of six "network backbones" that route all of mainland China's Internet traffic.

The NSA had also penetrated the Hong Kong headquarters of Pacnet, which operates one of the Asia-Pacific region's largest fiber-optic networks. Chinese state media had a field day accusing the U.S.

of hypocritically blaming them as villains for the same sort actions American's were perpetrating on China.

David Sanger reports that in almost every classified Pentagon scenario for how a future confrontation with Russia and China, even Iran and North Korea, might play out, the adversary's first strike against the United States would include a cyber-barrage aimed at civilians. It would fry power grids, stop trains, silence cell phones, and overwhelm the Internet.[332]

Cyber conflict remains in the gray area between war and peace, an uneasy equilibrium that often seems on the brink of spinning out of control. In the worst-case scenarios, food and water would begin to run out; hospitals would turn people away. Separated from their electronics, and thus their connections, Americans would panic, or turn against one another.

Black Holes in the Cyberspace

The enormous trove of documents leaked by NSA contractor Edward Snowden in 2013 revealed the incomprehensible capacity of advanced computing systems to collect, store, sift through, and decipher vast quantities of seemingly unimportant online metadata from sources across the world to construct a hidden "mosaic pictures" of information.

To be clear, "metadata" refers here to information that describes the nature of communication, rather than the content. In traditional telephone surveillance, for example, this would simply be a record of what phone number called another phone number at what time, as opposed to what was said on the call.

In the cyber era, metadata in the cyber era is far more expansive, including information about geographic location, time, email addresses, and other technical details about data being created or sent. When gathered together from sources around the world, sophisticated algorithms can be used to connect the dots and reveal new patterns, as well as to track individual devices and their users.

AI-enabled metadata collection and processing of an entire

haystack of data in order to find the proverbial needle of desired information was primarily developed by the NSA and other intelligence organizations to find bad guys. As with other advanced technologies, it can and is also used by cyberspace aliens against their pursuers and prey.

The term "cyberspace" used here broadly refers to every computer network in the world along with everything they connect to and control. These connections include the vast open network of networks known as the "Internet"—plus many other theoretically private and separate networks of computers that are *not* supposed to be accessible from the Internet.

Some parts of cyberspace are the "transactional networks" that do things like sending data about money flows, stock market trades, and credit card transactions.

Other networks are "control systems" that just allow machines to speak to other machines, like panels talking to pumps, elevators, and generators.

Under most conventional conditions these enormously complicated, interconnected networks function amazingly well. They do, at least, until cyberspace agents skyjack those network controls or crash them.

As previously discussed, the interconnectedness and interdependencies of these networks present an astounding abundance of cybersecurity vulnerabilities.

In their book *Cyber War,* Richard Clarke and Robert Knake discuss three general types of key vulnerability factors that concern cybersecurity professionals the most Internet design flaws; Internet software and hardware flaws; and an expanding number of critical online systems. [333]

Internet Design flaws:
A major Internet vulnerability resulted from the fact that it is one big network with a decentralized design. The designers did not want it to be controlled by governments, either singly or collectively, and

so they designed a system that placed a higher priority on decentralization than security.

While the design protocols were developed based upon rules which allowed for the massive growth in networking and the creation of the Internet as we know it today, the writers of its governing rules did not imagine that anyone other than well-meaning academics and government scientists would use it.

Much like the tribal areas of Pakistan or the tri-border region in South America, the Internet is not under the control of anyone. This has created countless cyber gap hideouts where many lawless culprits freely roam.

To illustrate a potential series of cyber disruption gaps, consider a "generic" network communication sequence.

First, the message being sent must identify which Internet Service Provider (ISP) will serve as the appropriate "carrier." Such ISP addressing information typically includes Internet access (e.g., email and World Wide Web); Internet transit (e.g., ability to transfer from a smaller network to a larger one); domain name registration (e.g., administrative authority); web hosting (e.g., offering World Wide Web access; Usenet service (e.g., enabling transfers to various topical "newsgroups"; and colocation center (e.g., data center or "carrier hotel.")

Some U.S. nation-wide ISPs own and operate thousands of miles of coast-to-coast fiber optic cables that connect all big cities. The six major ISPs (Tier 1 "backbone providers") are Verizon, AT&T, Quest, Sprint, Level 3, and Global Crossing.

Opening a browser on a laptop sends a request to go out into the internet and bring back the user's "homepage" according to the assigned "Domain Name System" on the computer server which is often referred to as the "web address."

A registered domain name is unique so that it can only be used to a single person or entity. It functions on the Internet in a similar way as a street address in the physical world...or at least that was the idea. Hackers frequently alter reputable domain name information and misdirect users to phony web pages that can appear to be entirely authentic. The strategy is to get the user to enter their secret account number and password.

Hackers also attack wireless "wi-fi" communications that provide access to a local area network.

Opening a browser sends a request to the server hosting the page which is broken down into a series of packets that are each sent individually. Of these separate packets, the first hop of one might be from the computer to the wi-fi card on the computer, where it is translated into radio waves and sent out over the air via a wi-fi router.

The wi-fi router then turns the signal back from a radio wave into an electronic signal passed to the user's local ISP. If the wi-fi router is improperly secured, hackers can get into the computer over the wi-fi connection.

Cyber warriors or criminals have at least two opportunities to intercept those packets and send them to the wrong place—or prevent them from going anywhere.

One option is to attack the Domain Name System, and as previously mentioned, route the user to a phony look-alike webpage.

Alternatively, rather than hacking the Domain Name System to hijack a webpage request, a cyber warrior or criminal can attack the system itself (targeting top-level worldwide domain servers with a DDoS attack.) This is what happened when botnets hit Estonia and Georgia.[334]

Hackers also attack major traffic regulation Border Gateway Protocol (BGP) system locations where ISPs come together—where one starts and another stops at a border. Information packets that cross the borders have routing labels with a "to" and "from" address, where BGP serves like a postal worker who decides what sorting station each packet goes to next.

BGP also does the job of establishing "peer" relationships between routers on two different networks. For example, going from AT&T to a lower Level 3 requires a BGP connection.

A big vulnerability is that the BGP system works on trust. A rogue insider working for a big ISP can hack into the BGP tables and spoof delivery instructions so that traffic becomes lost and doesn't reach its intended destination.[335]

Richard Clarke observes that just about everyone involved in network management for the big ISPs knows about these Domain Name System and BGP vulnerabilities.

Although parts of the U.S. government are deploying a secure Domain Name System, in practice, security protections are nearly nonexistent in the commercial infrastructure. System decisions are made by a nongovernmental international organization called the Internet Corporation for Assigned Names and Numbers (ICANN) which has little authority for overall Internet administrative guidance or control.

Still another vulnerability inherent in the Internet design is that almost everything that makes it work is open and unencrypted.

Much like a commercial radio scanner can listen in on two-way communication between truckers, and in most cities, between police personnel, the Internet generally works the same way.

In some cities, however, the police will "scramble" the signal so that criminal gangs cannot monitor police communications. Only someone with a radio that can unencrypt traffic can hear what is being said. To everyone else, it just sounds like static.

Only a fraction of Internet traffic is encrypted. So while it is a little more difficult to tune into someone else's messaging, it most certainly occurs. ISPs have access, and can give it to the government. Mail service providers like Google's Gmail also have access ...even if they claim that they don't.[336]

Third-party eavesdroppers "snoop" on Internet traffic information using a "pocket sniffer." This is an Internet wiretap device that can be installed on any operating system to steal other people's communications on a local area or Ethernet network.[337]

Although the standard Ethernet protocol tells a computer to ignore everything that is not addressed to it, this doesn't mean that it has to. An advanced Ethernet packet sniffer can look at all traffic. This means that one of your neighbors could sniff all Internet traffic on your street.

More advanced sniffers can trick the network in what is known as a "man-in-the-middle" attack where the sniffer appears to be the router as the user's computer. This way, all information is sent to the sniffer, which then copies it before passing it on to the real addressee.

Even though many (but not most) websites now use a secure, encrypted connection when you log on so that your password is not sent in the clear for anyone sniffing around to pick up, that still doesn't mean that your information is safe.

If a snoop wants to target someone, they can install a keystroke logger (either software or hardware) on their computer that secretly captures and transmits everything typed on the keyboard. This might happen, for example, if someone visits an infected website or downloads a phishing email file.

Internet software and hardware flaws:
Internet-connected equipment and users are subject to constant hacking scams and assaults from cybercriminals who take advantage of the system's ability to propagate intentionally malicious malware. In doing so they exploit software flaws, user errors like going to an infected site, and potentially—on a very large state government-sponsored scale—even compromise the hardware device manufacture or assembly.

As previously discussed, malware software tool called "Eternal Blue" developed by the NSA and later stolen by the North Korean Shadow Brokers group exploited a vulnerability in widely distributed Microsoft Windows to hack the Bangladesh Central Bank in 2016.

NSA never informed Microsoft about the vulnerability until after the Korean weapon based on it was stolen. And while the U.S. government says it reports to industry more than 90 percent of the software flaws it discovers, so that they can be fixed, Eternal Blue was a part of the 10 percent it had held on to bolster American firepower.[338]

In that case, the North Korean hackers married NSA's tool to a new form of ransomware that locked computers and made data inaccessible unless targeted users paid $300 for an electronic key. It is unclear how many paid, but those who did, never received the key—if there ever was one—to unlock their documents and databases.

Although banks and other systems across dozens of countries were affected, it is doubtful the North Koreans either knew or cared.

David Sanger points out that, in what some might see as a sign of cosmic digital justice, Russia's Interior Ministry was among the most prominent victims. The attack had been spread via a basic phishing email similar to the one used by Russian hackers in their attack on the Democratic National Committee and other targets in 2016.[339]

Briefly summarized again, viruses are programs passed from user to user (over the Internet or via a portable format like a flash drive) that carry some form of payload to either disrupt a computer's normal operation, provide a hidden access point to the system, or copy and steal private information.

Worms don't require a user to pass the program on to another user, Instead, they can copy themselves by taking advantage of known software code vulnerabilities, and "worm" their way across the Internet.

Phishing scams try to trick an Internet user into providing information such as bank account numbers and access codes by creating email messages and phony websites that pretend to be related to legitimate businesses.

Software vulnerabilities occur for a variety of reasons. For example, people who write long codes, like most all humans, make simple mistakes or get sloppy.

Those who plan malware defenses dynamically simulate suspicious computer code operation to determine whether a file examined will misbehave without putting the actual system at risk...just as a bomb squad rests suspicious packages, virtual "detonation chambers" can cause an incoming piece of malware to mistakenly think it is inside the machine and detonate prematurely.

A key software design and assessment factor are to recognize that most computer systems treat data both as information to be processed and commands to be executed. This fundamental principle is also a major source of insecurity.

One of the most common ways websites are attacked compromises a type of programming language used to manage data called "Structured Query Language" (SQL) which dates back to the 1970s.

Instead of entering a name and address as requested, a web server attack using SQL (pronounced "sequel") injection can enter specially crafted rogue commands that the databases read as program code rather than simply as data to be stored. These altered commands can then be used to learn the database, read data, create new accounts, and gain access to other files or permissions on the server.

In some cases, SQL injection can be used to discover and change the security settings on the server, allowing the attacker to control the entire web system. The Anonymous group used this kind of attack to penetrate the security firm HBGary and share its embarrassing secrets with the world.

Expanding the number of critical online systems:
The Internet is adding new connected systems and users at a rate that would have been incomprehensible a short time ago. In fact, the networks that aren't even supposed to be connected to the public Internet very often actually are, without their users' awareness.

As I discuss in my recent book The Weaponization of AI and the Internet: *How Global Networks of Infotech Overlords Are*

Expanding Their Control Over Our Lives, we are all subject to the ubiquitous Internet of Things privacy intrusions.

Most people likely don't know that copying machines are connected to the Internet can be hacked for information, or that paper shredders can be implanted with tiny cameras to secretly image and store information just before sensitive documents pass through the knives that cut them into little pieces. A cleaning crew guy can later collect and pass along the pictures to whoever hired him.

When a copier "phones home," it is doing exactly what the software is supposed to make it do. It is a very different matter entirely, however, if your competitor has a computer programmer who wrote a few lines and slipped them into the processor that runs your photocopier instructing it to store an image of everything and put them into a compressed data (or zip) file.

Then, perhaps daily, let's imagine that the copier accesses the Internet and—ping!—shoots the zip file across the country to your competitor.

That's not "supposed" to happen.

Office managers don't typically pay much attention when the salesperson tells them that the copier they are ordering will have a remote diagnostics capability to download improvements, fix problems, and dispatch a repairman with the right replacement parts.

Hackers, however, may have paid attention. Or maybe they were just exploring their cyberspace neighborhood and found an address that identified itself as "Xeonera Copier 2222, serial number 20-003488, at Your Company, Inc.

With the introduction of the web and wireless networks, all connected devices have an address. Accordingly, a diagnostic error report on a copier can be dispatched by the device itself even before a customer knows there is a problem.

Virtually every printer manufactured either has its own remote diagnostic tool (i.e. Ricoh's Remote, Kyocera Admin, Sharp's Admin,

Xerox's DRM) or have partnered with third party companies like Imaging Portals or Print Fleet.[340]

Independent cybercriminals and government state-sponsored hackers routinely use a few lines of weaponized software transported throughout the Internet of Things as a convenient avenue to take control of almost all connected devices. The first attack stage is often to obtain information, then to convert the operating system into a slave, and finally to add commands to be issued remotely.

Originally, the term "hacker" meant just somebody who could write instructions in code that is in the language of computers to get them to do new things. When they do something like going where they are not authorized, hackers become cybercriminals. When they work for the U.S. military, we call them cyber warriors.

Most concerning, early designers of electric power grids and other critical systems didn't think about such people—whether terrorists or "patriots"—hacking them and turning their systems into weapons.

The Pentagon even envisions a near-future scenario in which every soldier on the battlefield will be a hub in a network and as many as a dozen devices carried by that soldier will be plugged into the network and require their addresses.

As Richard Clarke, former White House National Coordinator for Security, Infrastructure Protection, and Counterterrorism observes:

> *What this could mean in a real-world conflict is something that until recently most policymakers in the Pentagon were loath to think about. It means that if you can hack into things on the Internet, you might not just be able to steal money. You might be able to cause some real damage, including damage to our military.*[341]

Nevertheless, no modern military can live without ever-smarter cyber capabilities, just as no nation could imagine, after 1918, living without airpower. Now again, just as then, it is impossible to imagine fully how dramatically this invention will alter the exercise of national power.

And thanks to rapid 5-G enhanced AI machine learning advancements, the invention will continue to reinvent itself.

5G Invites Malware Monsters

You can expect to witness numerous "telephone company" television commercials and printed news and magazine ads trumpeting the arrival of a "5G information revolution," along with the appearance of more and more new gray boxes on your street.

Some of those boxes will be attached to streetlights. Other larger ones mounted on streets will look much like those used by letter carriers to pick up mail. However, instead of belonging to the U.S. Postal Service, they will be owned by those same "telephone companies."

Those ubiquitous gray boxes we can expect soon to dot U.S. cities will contain the 5G transformers owned and operated by Verizon, AT&T, Sprint, and other carriers. The reason there will be so many of them is that, 5G waves don't travel very far—unlike 4G systems they will replace which can be mounted on high-up towers.

Nevertheless, what they lack in wave distance, they more than makeup for in processing capacity. This fifth generation of mobile telephony won't be twice as fast as the current 4G. It may be ten times as fast.[342]

Ultra-high-speed 5G networks in combination with Internet of Things connectedness ranging from medical-wearable devices to smart home appliances, to self-driving cars will continue to drive national, business and personal information security challenges at an unfathomable pace.

5G will supercharge the Internet of Things, while at the same time make these combined "information revolutions" both more awesomely influential and vulnerably insecure.

When 5G arrives near you, so will a new set of cyber risks along with it.

In a world of interconnected IoT networks, devices, and applications, every activity is a potential attack vector.[343]

Hackers have demonstrated abilities to compromise already vulnerable IoT devices with massive Distributed Denial of Service (DDoS) attacks dating back to 2016.

Consider now that the Russian 2017 NotPetya attack targeting Ukraine caused an estimated $10 billion in corporate losses; with combined losses at Merck, Maersk, and FedEx alone exceeding $1 billion. This occurred before 5G existed to greatly multiply the threat and damage level.

Given that operations of IoT devices heavily depend upon the speed and accessibility of the Internet, the next-generation 5G will also mean that it will take less and less time for hackers to overwhelm networks due to the exponential rate of traffic directed to targets.

The world's hackers (good and bad) are already turning to the 5G ecosystem, as the just concluded DEFCON 2019 (the annual ethical "hacker Olympics") illustrated.

The 2019 targets of hacker villages included key parts of the 5G ecosystem such as: aviation, automobiles, infrastructure control systems, privacy, retail call centers, help desks, hardware in general, drones, IoT, and voting machines.

Former FCC chairman Tom Wheeler and Rear Admiral David Simpson, USN (Ret.), former Chief of the FCC's Public Safety and Homeland Security Bureau, point out that next-generation 5G security protections will require a major overhaul of essential networks that will have long-term impacts. As they explain in a September 2019 Brookings Institute report:

> *Because 5G is the conversion to a mostly all-software network, future upgrades will be software updates much like the current upgrades to your*

smartphone. Because of the cyber vulnerabilities of software, the tougher part of the real 5G 'race' is to retool how we secure the most important network of the 21st century and the ecosystem of devices and applications that sprout from that network.[344]

Commenting on the urgency to spur 5G growth in the US, President Trump announced on April 12th 2019, that "The race to 5G is on, and America must win."

Since the software-based 5G network built on a distributed architecture precludes the kind of centralized malware intervention chokepoints afforded by earlier networks, 5G will afford virtual open invitations to attacks. Accordingly, primary responsibilities for 5G cyber protection must begin with companies who are actively involved in developing and implementing the new network, its devices, and applications.

Wheeler and Simpson emphasize that proactive responsibility for security of 5G services must address five key cyberattack vulnerabilities that differ with previous hardware-intensive network architectures:[345]

> *The 5G network has moved away from centralized, hardware-based switching to distributed, software-defined digital routing.*
>
> *Previous networks were hub-and-spoke designs in which everything came to hardware choke points where cyber hygiene could be practiced. In the 5G software-defined network, however, that activity is pushed outward to a web of digital routers throughout the network, thus denying the potential for choke point inspection and control.*
>
> *5G further complicates its cyber vulnerability by virtualizing in software higher-level network functions formerly performed by physical*

appliances. These activities are based on the common language of Internet Protocol and well-known operating systems.

Whether used by nation-states or criminal actors, these standardized building block protocols and systems have proven to be valuable tools for those seeking to do ill.

Even if it were possible to lock down the software vulnerabilities within the network, the network is also being managed by software—often early generation artificial intelligence—that itself can be vulnerable. An attacker that gains control of the software managing the networks can also control the network.

The dramatic expansion of bandwidth that makes 5G possible creates additional avenues of attack. Physically, low-cost, short-range, small-cell antennas deployed throughout urban areas become new hard targets.

Functionally, these cell sites will use 5G's Dynamic Spectrum Sharing capability in which multiple streams of information share the bandwidth in so-called "slices"—each slice with its own varying degree of cyber risk. When software allows the functions of the network to shift dynamically, cyber protection must also be dynamic rather than relying on a uniform lowest common denominator solution.

Finally, of course, is the vulnerability created by attaching tens of billions of hackable smart devices (actually, little computers) to the network colloquially referred to as IoT.

Plans are underway for a diverse. And a seemingly inexhaustible list of IoT-enabled activities, ranging from public safety things to battlefield things, to medical things, to transportation things—all of which are both wonderful and uniquely vulnerable. In July, for instance, Microsoft reported that Russian hackers had penetrated run-of-the-mill IoT devices to gain access to networks. From there, hackers discovered further insecure IoT devices into which they could plant exploitation software.

In November 2019, the National Security Telecommunications Advisory Committee (NSTAC)—composed of leaders in the telecommunications industry—told President Trump, "The cybersecurity threat now poses an existential threat to the future of the [n]ation."

Wheeler and Simpson warn that The 5G cybersecurity threat is a whole-of-the-nation peril.

Whereas early generation cyberattacks targeted intellectual property, extortion, and hacked databases, the stakes are even higher today as nation-state actors and their proxies gain footholds in our nation's critical infrastructure to create attack platforms.

Former Marine Corps Commandant, General Robert Neller told an audience in February 2019:

> *If you're asking me if I think we're at war, I think I'd say yes. We're at war right now in cyberspace. … They're pouring over the castle walls every day.*

General Neller must be referring in particular to cyber offenses deployed from within the Great Wall of China.

Huawei's Menace and Maneuvering

The U.S. government has prudently noticed that a lot of 5G equipment that is planned to be installed around the U.S. and the world will most likely be produced by a Chinese Internet electronics company named Huawei. The Pentagon and the National Security Agency never trusted that Chinese Internet electronics company, thinking that its products might be laced with Chinese government back doors and bugs.[346]

Most realistically, Huawei is not a private company or even a company. Rather, as described by the *Washington Times* national security columnist Bill Gertz, It is a deception operation masquerading as a company that until 2019 successfully fooled thousands of people around the world. The real purpose of Huawei, as with other faux companies, is industrial espionage, intellectual property theft, and human intelligence-gathering operations.[347]

This fact became clear in January 2019 when the Justice Department announced indictments against Huawei and its CEO, Meng Wanzhou. Huawei Technologies and two subsidiaries were charged with stealing robotics secrets from T-Mobile and illegally doing business with Iran.

Federal officials laid out the case for the first time with indictments asserting that Huawei used a front company in Iran, Skycom Tech Co. Ltd., to hide financial transactions with Iran worth tens of millions of dollars in a bid to avoid U.S. sanctions laws that are aimed at curbing financial dealings by the Islamist regime in Iran.[348]

In October 2019, the European Union released a report warning about a series of specific global security threats posed to 5G mobile networks by foreign state-backed vendors of telecommunications equipment, Huawei in particular.[349]

Huawei has been a major supplier of network gear in large European economies such as the UK and Germany. As the world's largest telecom equipment maker with 30% of the marketplace, the company leads other large competitors, Nokia Corp. of Finland, and Ericsson AB of Sweden.

Separately from that public October report, EU member states circulated a draft nonpublic risk analysis that highlighted specific security threats posed by telecom equipment suppliers from countries with "no democratic and legal restrictions in place." Once again, that assessment raised alarm among officials in European capitals concerning Huawei in particular.[350]

The EU analysis concluded that several member states have identified specific techniques that could be used in attacks, including the possibility that a vendor could insert concealed hardware, malicious software and software flaws into the 5G network.

As reported by the *Wall Street Journal—London Office*, an EU spokeswoman stated that the telecom security risks identified "may be related to the characteristics of individual suppliers, coupled with their particular role and involvement in 5G networks."

The U.S. and Australian government officials have long warned EU allies about national security risks posed by Huawei.

A particularly worrisome scenario would be Huawei installing deeply implanted flaws in the 5G network, providing Chinese intelligence services a menu of vulnerabilities to exploit on behalf of their government to spy on or disrupt all vulnerable networks.

Whereas the Chinese government has said it asks Chinese companies to follow local laws in international markets where they operate, critics argue that Chinese 5G technology companies are obliged to assist their government under the Chinese National Intelligence Law.[351]

The U.S. Federal Communications Commission Director Ajit Pai warns that allowing Chinese equipment into tomorrow's 5G wireless networks would open the door to censorship, surveillance, espionage, and other harms. The fact that Chinese law requires all companies subject to its jurisdiction to comply with requests from the country's intelligence services and to keep them secret means that China could compel Huawei to spy on American individuals and businesses. Writing in the Wall Street Journal, Pai projects:

Imagine if a 5G network with Huawei equipment were operating near a U.S. military installation, critical infrastructure facility or other sensitive location. Beijing could demand the installation of a 'back door' to allow secret access to the network, insert malware or viruses, and receive all kinds of information—without Americans ever knowing.[352]

The FCC Director states that independent experts confirm the risk:

A report issued this year [2019] by cybersecurity firm Finite State found a majority of Huawei firmware images it analyzed had at least one potential back door and that each Huawei device had an average of 102 known vulnerabilities.[353]

Commissioner Pai adds:

We also need to make sure existing networks are secure. Some rural wireless carriers that receive money from the [annual FCC Universal Service] Fund have already installed Chinese equipment. That poses an unacceptable risk.[354]

Escalating tensions between China and the U.S. along with other Western nations have further raised concerns regarding cyber threats posed by Huawei, ZTE, and other Beijing government-subordinated technology companies.

Given these very real concerns, the U.S. government has launched a concerted campaign both domestically and internationally to block Huawei from building Next-Generation (5G) wireless networks.

The U.S. government and corporate organizations also recognize economic threats posed to American telecommunication product markets by Chinese government-subsidized and legislatively-advantaged information technology companies.

Huawei's cheaper product costs and global market penetrations are believed to be a factor causing U.S. technology firms to lag behind some other countries in the global race to roll out 5G technology. This circumstance, added to national security vulnerabilities, prompted the 2019 Trump administration executive order bans on Huawei and other Chinese-manufactured products entering the U.S. market.[355]

Previously, in 2018, the White House had blocked Singapore-based Broadcom's proposed $105 billion acquisition of Qualcomm, a U.S. semiconductor and telecommunications equipment company, citing concerns about the deal's effect on America's ability to compete with China in 5G technology.[356]

Writing in *Georgetown Security Study Review*, Daniel Zhang argues that the Trump administration also invoked national security concerns to block Broadcom's bid out of concern that the Chinese company's real objective was to undermine Qualcomm's ability to innovate and compete in the emerging global 5G marketplace.[357]

In any case, Zhang points out that focusing exclusively on preventing Chinese hardware from supporting U.S. networks would create a false sense of security that ignores the risks to other parts of the next-generation network, namely the software and applications.

Former U.S. FCC Chair Tom Wheeler and Brookings Institution colleague Brookings Institution. Rear Admiral David Simpson, USN (Ret.) points out that China is a threat even when there is no Huawei equipment in our networks. Writing in a September 2019 Brookings report, they observe:

> *From the successful exfiltration of highly sensitive security clearance data in the Office of Personnel Management breach commonly attributed to China, to the ongoing China-linked threat actor campaign against managed service providers, many of China's most successful attacks have taken advantage of vulnerabilities in non-Chinese applications and*

hardware and poor cyber hygiene. None of this goes away with the ban on Huawei.[358]

Wheeler and Simpson emphasize that individual companies must recognize and be held responsible for protecting their systems and use:

> *There needs to be a new corporate culture in which cyber risk is treated as an essential corporate duty and rewarded with appropriate incentives, whether in monetary, regulatory, or other forms. Such incentives would require adherence to a standard of cyber hygiene which, if met, would entitle the company to be treated differently than other non-complying entities.*[359]

The two Brookings Institution authors further clarify that since future 5G cyberattacks will be software attacks, they must be countered by software protections. They report:

> *During a Brookings-convened discussion on 5G cybersecurity, one participant observed, 'We're fighting a software fight with people,' whereas the attackers are machines.*
>
> *Such an approach was like 'looking through soda straws at separate, discrete portions of the environment' at a time when a holistic approach and consistent visibility across the entire environment is needed. The speed and breadth of computer-driven cyber attacks require the speed and breadth of computer-driven protections at all levels of the supply chain.*[360]

Tom Wheeler and David Simpson portray an urgent but

discouraging outlook at American preparedness to even assess, much less address, escalating 5G network cybersecurity threats. Whereas placing greater security burdens upon the corporate sector is vital, accomplishing this will require major transformational business priority changes:

> *Looking to the cybersecurity roles of the multitude of companies in the 5G "ecosystem of ecosystems" reveals an undefined mush. Our present trajectory will not close the cyber gap as 5G greatly expands both the number of connected devices and the categories of activities relying on 5G.*
>
> *This general dissonance is further exacerbated by positioning Chinese technological infection of U.S. critical infrastructure as the essential cyber challenge before us. The truth is that it's just one of many.*[361]

Wheeler and Simpson point out, for example, that corporate equipment and software purchasers must be careful to check the origins of their digital supply chain sources. By and large, that isn't happening.

In 2016, hackers shut down major portions of the internet by taking control of millions of low-cost chips in the motherboards of video security cameras and digital video recorders. Such security vulnerabilities result when companies give little thought to cybersecurity in opting for low-cost connected devices, and retailers don't prioritize security in decisions of what to stock.[362]

Keeping Chinese hardware out of the U.S. 5G network does not equate to successfully preventing foreign cyber threats. Regardless of who builds the American 5G network, there will be cybersecurity risks. Hardware, like that offered by Huawei, is but one part of 5G.

Daniel Zhang notes in his June 7th 2019, *Georgetown Security*

Studies Review article:

> Both Russia and North Korea have successfully
> infiltrated networks in the U.S. without exploiting
> Chinese hardware. So, to effectively protect the
> security of its 5G networks, the U.S. should
> conduct a national risk assessment of 5G
> infrastructures on a larger scale, assessing if the
> current security requirements for network
> providers are sufficient to ensure the security of
> next-generation networks.[363]

Wheeler and Simpson concluded that the adage, "what's everybody's business is nobody's business," has never been more appropriate—and dangerous—than in the quest for 5G cybersecurity:

> Never have the essential networks and services that
> define our lives, our economy, and our national
> security had so many participants, each reliant on
> the other—and none of which have the final
> responsibility for cybersecurity.
>
> The after-the-fact cost of missing a proactive
> 5G cybersecurity opportunity will be much greater
> than the cost of cyber diligence up front.
>
> The time to address these issues is now, before
> we become dependent on insecure 5G services with
> no plan for how we sustain cyber readiness for the
> larger 5G ecosystem.
>
> The new capabilities made possible by new
> applications riding 5G networks hold tremendous
> promise. As we pursue the connected future,
> however, we must place equivalent—if not
> greater—focus on the security of those
> connections, devices, and applications.

They stress that to build 5G on top of a weak cybersecurity foundation is to build on sand. This is not just a matter of the safety of network users, it is a matter of national security.

Yes, the 'race' to 5G is on—but it is a race to secure our nation, our economy, and our citizens.[364]

NSA Opens Back Doors to China

In his book *"The Perfect Weapon," New York Times* national security correspondent David Sanger reports on a covert NSA program approved by the Bush White House to do to Huawei what they suspected China was using the company to do to America and our Western allies.

Termed "Shotgiant," the plan was to bore a way deep into Huawei's hermetically sealed headquarters in Shenzhen, China, crawl through the company's networks, understand its vulnerabilities and tap the communications of its top executives.

That, however, was only the very beginning. Next, the strategy was to exploit Huawei's technology so that when the company sold equipment to other countries—including allies like South Korea and adversaries like Venezuela—the NSA could roam through those nations' networks.

As Sanger explains it, there was another goal as well to prove the American accusation that the PLA was, as they expected, secretly running Huawei and that the company was secretly doing the bidding of Chinese intelligence.[365]

The American concern about Huawei was justifiable. NSA believed that China had made concerted efforts to penetrate deep inside U.S. communication networks. Similar worries applied to the Kremlin's Kaspersky Lab, the Russian antivirus software maker whose products were making it easy for Russian intelligence agents to exfiltrate secret American documents.

At least that had been the Shotgiant plan until 2013 when the

German weekly *Der Spiegel* and the *New York Times* published the details based on Snowden documents. The public release of those secrets and revelations of obvious hypocrisy aroused widespread criticism both in China and among many American allies.

US officials who were willing to talk about Shotgiant did their best to justify the action, arguing that NSA only breaks into foreign networks only for "legitimate" national security purposes. Catlin Hayden, then the NSC's spokeswoman said:

> *We do not give intelligence we collect to U.S. companies to enhance their international competitiveness or increase their bottom line. Many countries cannot say the same.*[366]

Sanger points out a problem that the Chinese did not distinguish between "economic advantage" and "national security advantage." As Huawei invested in new technology and laid undersea cables to connect its networking empire, the agency was interested in tunneling into key Chinese customers, including "high priority targets—Iran, Afghanistan, Pakistan, Kenya, Cuba."

He writes:

> *In short, eager as the NSA was to figure out whether Huawei was the PLA's puppet, it was more interested in putting its own back doors into Huawei's networks. It was a particularly important mission because the Chinese firm was popular in hard-to-access countries where American telecommunications companies were unlikely to ever get a contract.*
>
> *In other words, Huawei might serve as a back door to the PLA, but it would also be host to another back door, one it didn't know about: to the NSA.*[367]

In addition to Huawei, The Snowden files revealed that in 2013 NSA also cracked two of China's biggest cell phone networks and was happy to discover that some of the most strategically important units of the Chinese Army including some that used the easy-to-track devices to maintain its nuclear weapon systems.

Still, other Snowden documents described how NSA had mapped where the Chinese leadership lived and worked. A huge bull's-eye had been placed on Zhongnanhai—the walled compound next to the Forbidden City that was once a playground of emperors and their concubines.

Conveniently, NSA had discovered that Chinese targets had ill-protected Wi-Fi networks, and like everyone else, were constantly complaining about how slow the service was…leading to demands to upgrade their equipment. That created an opportunity for the Tailored Access Operations unit to intervene.

Beginning around about 2008, the NSA began making use of new tools designed to steal or alter data in targeted computers even when they weren't connected to a network—just as they did with Israeli intelligence in Iran to get past the air gap that separated the Natanz plant from the digital world.

Snowden's revelations revealed a comprehensive "ANT" catalog which updated "bugs" that intelligence agencies had developed. The most ingenious of the devices relied on a covert channel of low-frequency radio waves transmitted from tiny circuit boards and USB keys inserted surreptitiously into the target computers.

David Sanger noted again that getting the equipment into the computers required, of course, that the United States or one of its allies insert the hardware into the devices before they were shipped from the factory, divert them while they were in transit, or find a stealthy spy with a way to gain access to them was no easy task.[368]

But sometimes it was also possible to fool a target into inserting the devices themselves.

Snowden's ANT catalog revealed a new class of hardware with a scale and sophistication that enabled NSA to get into—and alter data on—computers and networks that their operators thought were completely sealed off from the Internet, and thus impermeable to outside attack.

One such device called "Cottonmouth" looked like a normal USB plug that you might buy at Office Depot. But it had a tiny transceiver buried inside that leapt onto a covert radio channel to allow "data infiltration and exfiltration."

Once the illicit circuitry was in place, signals from the computer were sent to a briefcase-size relay station—wonderfully called "Nightstand"—that intelligence agencies could place up to eight miles away from the target.

The NSA had even gone to the trouble of setting up two data centers China through front companies whose main purpose was to insert malware into computer systems. To accomplish this, the NSA used an Internet exploitation system they called "QUANTUM."

The QUANTUM mechanism was also used beyond China, with parallel efforts to get malware into Russian military networks and systems used by the Mexican police and drug cartels.[369]

Following disclosure by the *New York Times*, NSA officials predictably declined to confirm, at least on record, that the Snowden documents described any of their programs. "Off the record messaging," however, asserted that it was all part of a new doctrine of "active defense" against foreign cyberattacks.

In other words, the activities were primarily aimed at surveillance rather than at "computer network attack"—NSA-speak for "offensive action."

David Sanger observed that NSA officials didn't believe that the ANT catalog came from any documents that Snowden's "crawler" worm had touched and began looking for another insider—a "second Snowden."

Adm. Michel Rogers who later took over as director of NSA and chief of U.S. cyber Command acknowledged to Sanger in 2014 that:

...many allies were angry—some because they discovered Washington was spying on them (like the Germans), and others because Snowden revealed that they were secretly helping the Americans (the list was long, but also included the Germans).[370]

Rogers' greatest concern, according to Sanger, was the need for leaders who were well aware of being spied upon would find a need to publicly condemn American overreach. This public posturing would ,therefore, have a corrosive effect on future cooperation.

Peter Singer and Alan Friedman at the Brookings Institution urge U.S. to remember:

Before we point too many fingers at China for operating in this realm, remember that the United States is just as active; indeed, large parts of its intelligence apparatus, like the CIA and NSA, are dedicated to the same mission; the 2013 Snowden leaks showed 213 of these operations in 2011 alone.[371]

This being the case, China is characterized by time and time again in the U.S. and other government reports as the most active and persistent bane of the cybersecurity world. The Pentagon's 2011 *"Strategy for Operating in Cyberspace"* designed to guide Cyber Command, for example, cited China among the most important U.S. threats in this realm.[372]

Chinese writers and officials assert their country is the aggrieved party... the one most frequently under attack.

Singer and Friedman confirm that in one sense—that of raw numbers of cyberattacks—they are right:

The Chinese Ministry of Public Security has

reported that the number of cyberattacks on Chinese computers and websites has soared by more than 80 percent annually, while some 10 to 19 million or more Chinese computers are estimated to be part of botnets controlled by foreign computer hackers.

Beijing officials claimed in 2011 that China was the target of some 34,000 cyberattacks from the US, while in 2012 the numbers escalated to the point that Chinese military sites alone were targeted by American sources almost 90,000 times. A large number of these attacks originate from within the US.

But the story really isn't that simple.

While China is right when it says that it is a victim of hacking, the main culprit is it's disregard for intellectual property, not state-sponsored espionage.

A key reason why China's heavy malware infection rate is because as much as 95 percent of the software that Chinese computers use is pirated. Accordingly, this means that it doesn't get the same security upgrades and patches that legal license holders do, leaving them vulnerable to basic threats.[373]

Quantum Computing Ups the Ante

A July 2015 China State Council report outlined a three-phase effort that will culminate in 2049—the one-hundredth anniversary of the founding of Communist China—when they intend for China to dominate the world's economy with advanced technology and industrial systems. Telecommunications is the key element of the strategy. Under "broadband penetration" of world markets, China plans to go from 37 percent of the global market to 82 percent by 2025.

As outlined in the strategy, China plans to:
Master core technologies like new computing, high-speed

Internet, advanced storage and systematic security.

Make breakthroughs in fifth-generation mobile communications (5G), core routing switching, super high-speed and large capacity intelligent optical transmission, and core technology and architecture of the future network.

Research equipment like high-end servers, mass storage, new routing switches, new intelligent terminals, next-generation base stations, and network security to promote systematization and scale applications of core communication equipment.

Promote quantum computing and neural networking.[374]

To achieve those goals, Huawei and every Chinese company will be cooperating closely with the PLA and Ministry of State Security and other organs of the Communist Party-state.

In 2016, China's 13th five-year plan identified quantum computing as a key strategic initiative and authorized a quantum "megaproject" to that end.

China's estimated tens of billions in quantum commitments far outpace Washington's more modest outlays.

Vinod Vaikuntanathan, Chief Cryptographer at Duality Technologies, observed:

> To put these numbers into perspective, the entire U.S. budget for quantum research in 2016 was just $200 million, or one-fifth the cost of a single Chinese project in the field.[375]

While the U.S. Trump administration signed the National Quantum Initiative Act in December 2019, providing $1.2 billion in quantum research funding over five years, Beijing's efforts to promote quantum technologies remain unmatched.

In 2016, China launched Micius, the world's first quantum communications enabled satellite to conduct a presently "unhackable" two-way encrypted video call. For some, that launch eerily echoed the launch of the Soviet Union's Sputnik satellite in

1957 which caught the United States off guard and spurred a decades-long contest to regain and maintain global technological and military supremacy.

Vaikuntanathan notes that the parallel wasn't lost on the Chinese. Jian-Wei Pan, the lead researcher on the Micius project, hailed the start of "a worldwide quantum space race."[376]

Vaikuntanathan observes that China's massive population and lax privacy regulations already give it an AI edge, as more people and fewer restrictions means more accessible data.

> *For the U.S. to compete—and protect its citizens' rights to privacy—future data encryption schemes must protect data not just during transmission and storage, but also during analysis. Lattice-based homomorphic encryption strikes an ideal balance between ensuring post-quantum security, while not sacrificing computational utility.*

Former senior U.S. State Department Antiterrorism Assistance Program advisor, Morgan Wright, compares China's current development of a new $10 billion 4 million square foot quantum computing park in the Anhui Province with Bletchley Park, home to Winston Churchill's massive World War II "Ultra" program to intercept and decode information coming from previously unbreakable German "Enigma" machines.[377]

Wright predicts that new generation quantum encryption will be a major cybersurveillance and security gamechanger:

> *There is no way to "eavesdrop" and listen in. No way to siphon off the message traffic by tapping the line.*
>
> *We will be completely blind. And with quantum encryption, our adversaries will absolutely know we're trying to listen in.*[378]

Does China have better scientists than the United States?
Wright believes not:

> *What they do have is money when they need it to solve problems. Right now. No messy Congressional oversight hearings. No public outcry about how taxpayer money is being spent. Unheard of in Communist China. (Or only heard once.)*
>
> *What they also have is a sophisticated, long-range, long-term operation in the United States to steal our intellectual property...including aggressively placed operatives at universities to include professors, scientists, and students.*
>
> *The advancements China has made have not as much to do with discovery as they do espionage and intellectual property theft. You can spend 20 years trying to invent something, or two years stealing all the data you need.*
>
> *Stealing is cheaper, faster and doesn't usually start wars.*[379]

And what makes this an urgent national security matter?

Because as Vinod Vaikuntanathan points out, this new computing paradigm will enable a quantum leap—pun intended—in processing power, whoever masters it will cement their supremacy across almost every key technological domain. Not only will viable quantum computers represent a landmark achievement, but they will also carry seismic geopolitical implications—especially in the critical domains of information security and cyberwarfare:

> *Everything from web traffic to e-Commerce to block chain relies on something called public-key cryptography, which allows users to encrypt data with a shared public key but decrypt it with their*

own distinct private key. The public and private keys are mathematically connected in a way that's easy to compute in one direction but almost impossible to reverse—almost impossible for conventional computers, at least. Using Shor's algorithm, quantum computers could crack these codes with ease.[380]

Threats posed by code breaking quantum computers loom large. Much as nuclear weapons upended the calculus of conventional warfare, capable quantum espionage and attack weapons will redraw the lines of cyberwarfare.

There will no longer be any such thing as a secure communication channel using the current encryption standard called "Advanced Encryption Standard" (AES) which forms the backbone of many of our secure messaging systems. Accordingly, the NSA has called for the phasing out of all vulnerable encryption systems and NIST is administering a nationwide contest to accelerate innovation in post-quantum cryptography.

But the big follow-on question remains; while quantum encryption is considered by some to be unbreakable now...in the new quantum era, is anything truly unbreakable?

The jury of history remains undecided on this matter.

At least for now, quantum-proof cryptographic systems currently do exist.

One such system is known as "lattice-based encryption" links Public and Private "keys" with the mathematics that are difficult to crack even for quantum computers. Each public and private key pair is comprised of two uniquely related cryptographic keys comprised of long strings of random numbers. As the name suggests, the Public key is available to anyone via a publicly accessible repository or directory, and the Private key must remain confidential to its respective user.

Lattice-based encryption offers an additional advantage in being fully "homomorphic," meaning that it allows computation on encrypted data without ever exposing the data itself.

Kiran Bhagotra, CEO and Founder of the cybersecurity firm ProtectBox, doubts the much-lauded "tamper-proof" claim of "blockchain" technology that underpins digital cryptocurrencies such as Bitcoin and is being integrated into everything from insurance to immigration. Instead, she suspects that the open and versatile nature which makes it particularly appealing may also create vulnerabilities to undoing by quantum computers.[381]

Writing in *Wired*, Bhagotra explains her reasoning that unlike traditional computers that solve a complex computational maze by going down one path at a time, quantum machines can go down multiple paths simultaneously. This feature which may make quantum systems thousands—or even millions—of times faster than current computers will also enable them to apply those powers to break encryptions massively faster and easier as well.

Blockchain would then be particularly vulnerable because it has a single point of failure—all the blocks are freely available for criminal organizations or surveillance operatives to gather.

In the meantime, people are putting processors in everything and connecting everything to networks at an amazing pace.

Bhagotra theorizes that when quantum computing gets cheap enough, there could be huge leaks of blockchain data that is potentially being stockpiled now:

> *It's like keeping your valuables in a padlocked box but then leaving it out on the street. Quantum's 'entanglement' ability also enables it to make logical leaps conventional computers never would. It could fill in the gaps between data harvested from just a few blocks.*[382]

Given the dire consequences of falling behind, several American governmental, institutional and corporate organizations are investing heavily in a race to develop quantum competitiveness and data protection.

In January 2019, IBM unveiled its latest quantum computer. At just 20 qubits, the IBM Q System One was impressive but far from revolutionary. Other American tech heavyweights like Google and Intel are funding similar research, and while the results are promising, "quantum supremacy" still lies beyond the horizon.[383]

Cybersecurity experts Richard Clarke and Robert Knake confirm that at least one new encryption method being explored uses a form of quantum computing to secure messages, using quantum to deal with quantum. They foresee in their book, The *Fifth Domain*, that such quantum supremacy, should it occur, could ultimately have a single victor, one who frighteningly conquers and controls all AI networks:

> *If [we are] to combine a truly operational quantum computer with some specialized machine learning and orchestration applications for running and defending a classical computer network, it might just be possible to deal with the millions of actions that are simultaneously taking place on a network and its periphery, taking into account all of the data that is in storage about the network, factoring in information about what is happening in near real time elsewhere in cyberspace, and repairing or writing code on the fly.*
>
> *In short, you might be able to create the 'one AI to rule them all' on a network. You might actually be able to defend a network successfully. Or attack one.*[384]

Former University of California President C.L. Max Nikias offered the prediction that "whoever gets this technology first will be able to cripple traditional defenses and power grids and manipulate the global economy." [385]

Clarke and Knake warn that while that Nikias prediction may prove overstated, given known tendencies of Russia, China, and the U.S. militaries and defense services, it probably will not take them long to start thinking like this. Militaries think first of offensive weapons. It's in their DNA.

Yet what if a one-winner-take-all quantum arms race becomes real?

Richard Clarke and Robert Knake hypothesize that if an attack algorithm were written in new quantum code, taking advantage of an operational quantum computer's computational capacity, it might be possible to develop a tool that would collect everything that is known about a network, simulate it, and find the best way to attack it:

> *Indeed, it could be possible to design a series of optimized attacks for a host of networks and then launch them more or less simultaneously, bringing a target nation or group of nations to a pre-industrial era condition in seconds or minutes.*
>
> *Such a crippling attack is probably too difficult for teams of humans to execute today, but an operational quantum computer with bespoke, optimized machine learning/quantum programming might be able to do it.* [386]

David Sanger agrees that there is good cause for worry that within just a few years these weapons, merged with artificial intelligence, will act with such hyperspeed that escalatory attacks will take place before humans have the time—or good sense—to intervene:

> *We keep digging for more technological solutions—bigger firewalls, better passwords, better detection systems—to build the equivalent*

*of France's Maginot Line. Adversaries do what
Germany did: they keep finding ways around the
line.*[387]

Meanwhile, Sanger states, our adversaries are always thinking
forward to a new era in which such walls pose no obstacle and cyber
is used to win conflicts before they appear to start.

*They look at quantum computers and see a
technology that could break any form of encryption
and perhaps get into the command-and-control
systems of America's nuclear arsenal. They look at
bots that could not only replicate real people on
Twitter but paralyze early-warning satellites.*[388]

Clarke and Knake urges U.S. to think of the quantum computing era
as a Y2K for encryption: a time when everyone is forced to update
their software to ensure that a hacker with access to a quantum
computer cannot become a problem. It is not a direct or immediate
threat quite yet, but it will arrive sooner than most realize.

Whether two years away or ten banks and other commercial
and private-sector organizations are going to have to shift from
encryption systems they use now to quantum-resistant systems.

Experts urge that active planning to transition to post-quantum
cryptography begin now. Although the timing of technology
breakthroughs is a matter of conjecture, adversarial intelligence
agencies may already be gathering troves of data to decrypt once
quantum technology arrives.

From NSA headquarters at Fort Meade to the national
laboratories that created the atomic bomb, American scientists and
engineers are, in fact, struggling to maintain a lead.

Each year researchers from governments and computing giants
including Intel, Microsoft, and Cisco meet with top academics in the
worlds of cybersecurity and mathematics at a "Workshop on

Quantum-Safe Cryptography" to discuss potential solutions to the common threat.

An enormous amount of collaborative government, academic and corporate work going on to marry quantum computing and machine learning. NASA, Stanford, and Google have come together to create QuAIL, the Quantum and AI Laboratory in Palo Alto. And even before there is a real operational quantum computer, teams at places such as MIT and the University of Toronto are busy writing machine learning applications in the new computer languages developed for quantum.[389]

Vinod Vaikuntanathan of Duality Technologies describes quantum computing is an emblematic new battleground. He predicts that mastering such cutting-edge technology will not only garner prestige, but it could also very well help determine the global balance of power:

> In short, quantum computing is this century's moonshot—and now (as then), its outcome is about far more than national pride. It's nothing less than a matter of national security.[390]

A NEW FIFTH DOMAIN OF CONFLICT

THE PENTAGON HAS traditionally presided over four domains of conflict: land, sea, air, and space. Cyber, now a fifth domain, is different in part because the risks fall outside these historical norms for a variety of important reasons:

Cyberattacks occur at digital speed:

A traditional defense system that cannot respond in "digital time" to a multi-pronged threat or that cannot provide protection while attacking others may be of little use in the future.

Warfighters must operate—and when necessary fight—seamlessly from undersea, surface, land, air, and cyber space at the speed of light.

Potential threats must be identified, properly attributed, assessed and mitigated with little or no time for after-the-fact analytics nor tolerance for errors.

Much of the military national defense focus will prioritize disrupting the opponent, while simultaneously avoiding escalation to lethal conflict situations.

Cyberwarfare has no territorial boundaries nor assured targeting barriers:

Someone in Brazil can compromise computers in South Africa to launch attacks on systems in China, which might be controlled by computers physically located in the US.

Although sophisticated analysis may identify the computer being used to launch the attack, it is far more difficult to determine whether that computer is being operated remotely and, if so, by whom.

In April 2007, NATO and its collective defense ideals faced a twenty-first-century test when Estonia, new alliance member, found itself under a furious DDoS cyber bombardment that knocked banks and government services offline.

While the attack resulted from a series of botnets that had captured over a million computers in seventy-five countries, Estonia quickly pointed the finger at neighboring Russia.

Of the computers that attacked Estonia in the cyber incidents of 2007, 25 percent were US-based, even though the attack was originally Russian sourced.

Cyber weapons can strike instantaneously and persistently with uncertain motivations posing as commercial rather than military agendas.

The 2014, North Korea attack on Sony Pictures Entertainment showed that not only government systems and data that could be targeted by state-backed hackers.

A federal grand jury in Atlanta indicted two Iranians for allegedly hacking into the city of Atlanta's network in March 2018. Atlanta refused to pay a ransom of $52,000 in bitcoin...but endured millions of dollars in losses from the attack.

Another March 2018 ransomware attack on Baltimore delayed home sales and prevented the city from issuing water bills. The malware which exploited a Microsoft software weakness had been stolen from NSA also infected systems in Texas and Pennsylvania

Back door malware inserted to infect software programs and

hardware equipment can operate undetected over long time-periods:

Many bugs carry malware infections deep into critical foreign communications and control networks that can later be crippled or destroyed when desired.

There are many ways of penetrating networks and assuming the role of a network administrator or other authorized users without ever doing anything that would cause an alarm.

Moreover, if an alarm does go off, it is often such a routine occurrence on a large network that nothing will happen in response because no one noticed that anything out of the ordinary happened.

Cyberattacks targeted on critical infrastructure such as energy grids can cause catastrophic damage and deaths in addition to enormous collateral damage.

A 2007 Idaho National Laboratory digital attack test proved that malware could be used to destroy physical objects—in that case, a generator.

The 2010 Stuxnet malware attack on an Iranian nuclear weapon production plant reportedly ruined nearly one- fifth of the country's nuclear centrifuges.

Just before Christmas in 2015, hackers managed to disrupt the power supply in parts of Ukraine, by using a well-known Trojan called Black Energy.

In March 2016, seven Iranian hackers were accused of trying to shut down a New York dam in a federal grand jury indictment.

The most "wired" countries and organizations are most vulnerable through the Internet of Things.

Cyberwarfare also will evolve as the Internet evolves, including the growth of digitally-enabled appliances and everyday items.

Similarly, an "Internet of Military Things," which arises from increasing connectivity in aircraft, weapons, air defense, and communications systems, and personal protective equipment opens the U.S. military up to a new range of weapons, opponents, and threats.

Paradoxically, the same cyber tools that will advance U.S.

military capabilities are also creating additional vulnerabilities because they operate on the same Internet that is under attack. This creates a necessity either for the military and critical infrastructure systems to create a new, independent and secure Internet, or to ensure that their systems can be reliably insulated or encrypted in some presently unknown way from the open Internet.

The cyberwarfare era also presents an entirely different sort of battlespace than previous times when combatants always knew who they were fighting, could more confidently anticipate and prepare in advance for their opponents' strike capabilities and actions and could intimidate and dominate enemy forces through the benefit of superior troop strength and physical armaments.

No longer do powerful aggressors and defenders need to depend upon costly weapons of mass-destruction devices and sophisticated intercontinental missile delivery systems. Instead, global cyberwarfare is fought on keyboards by armies of ones and zeros...or at they will be until such time that quantum computers make those "old" binary programs pitifully obsolete.

Even the traditional definition of what constitutes and rises to the level of being termed "warfare" has changed in ways that remain open to dispute by different "experts."

Peter Singer and Alan Friedman emphasize a common conceptual disconnect between an actual state of war and far more frequent uses and misuses of a term like "cyberwar:"

> *War is used to describe an enormously diverse set of conditions and behaviors, from a state of armed conflict between nations (World War II) to symbolic contestations (New York City's "war on sugar").*
>
> *As for "cyberwar", the term has been used to describe everything from a campaign of cyber vandalism and disruption (the "Russian Estonian cyberwar", as it is too often called), to an actual*

state of warfare utilizing cyber means.[391]

Whereas cyberwar has been portrayed as everything from military conflicts to credit card fraud, the U.S. government's position defines a cyberattack as something violent that would "proximately result in death, injury or significant destruction." And whether waged on land, at sea and/or in space, all warfare has generally been distinguished from criminal enterprise by a broad nation-state or political regime agendas.[392]

A Former head of counterintelligence under the U.S. Director of National Intelligence Joel Brenner perceives that cyberwar further blurs fuzzy lines between periods of conflict and peace:

"We in the U.S. tend to think of war and peace as an on-off toggle switch—either at full-scale war or enjoying peace" says Joel Brenner, former head of counterintelligence under the U.S. Director of National Intelligence.

> *The reality is different. We are now in a constant state of conflict among nations that rarely gets to open warfare...What we have to get used to is that even countries like China, with which we are certainly not at war, are in intensive cyberconflict with us.*[393]

If espionage and covert placement of devastatingly destructive weapons in critical infrastructure systems such as electrical power grids and vital communication networks qualify, then we—the world—are already at war.

If "cyberwarfare" requires a national political agenda, as some authorities cited in this book suggest, then what about cyber theft of proprietary commercial secrets and classified strategic technologies? Would grand-scale for-profit ransomware extortions against companies and communities by criminal enterprises be included? And do cyber terrorism acts by rogue groups—large or small—

qualify as "warfare?"

Whether or not any or all of these various threats are directly attributable to nation state-sponsored, their proxies or entirely independent agents ultimately matters little to the targeted or collateral victims…a group that potentially includes all of us.

Many cyber weapons are based on software vulnerabilities and those vulnerabilities exist on numerous systems. Since the malware isn't always surgically-targeted, the potential of harming those not even involved is great. The assaults can affect networks that drive health care, manufacturing, power generation and distribution, and transportation, among others.

Once an attack tool is used, rather than being spent like a round of ammunition, it can be reused against ever more geographically far-ranging and diverse targets.

Former President George W. Bush Special Assistant for Homeland Security, Marie O'Neill Sciarrone, explains why traditionally targeted solutions no longer apply. Writing in a 2017 George W. Bush institute report, she observes:

> *A new paradigm must be developed that reflects the realities of cyberspace, which expands the battlefield anywhere to which the internet extends, particularly past the supposedly-safe borders of our homeland and into almost every aspect of our lives. That so much of business, political, and social activity relies almost exclusively on this technology means escaping the impact of cyber warfare is unlikely. The capacity for a single solution is equally unlikely.*
>
> *What's more, definitive attribution of the adversarial act(s) can be difficult or even impossible. A single person can control an army of usually-unwitting computers, making it even more difficult to identify who is behind the actions.*[394]

Sciarrone points out that attribution difficulties in an era when a single individual can take control of an army of zombie computers add significant defense policy challenges. Unlike traditional warfare, it becomes unclear regarding who has what authority to respond in a significant cyberattack when they respond, and what options leadership can enact.

As intelligence gathering continues to expand and pervade nearly every sphere of influence, it becomes difficult or impossible to distinguish between digital intrusions aimed at routine government intelligence collection versus attempted interceptions or interruptions that signal plans for an attack.

An enemy, for example, may disrupt a defense system while inserting malicious code to collect information from our systems as part of traditional espionage information-gathering. That same malware also could be intentionally inserted to disrupt and take down the system for more nefarious purposes.[395]

Marie O'Neill Sciarrone maintains that the capacity to determine the difference—or where an exploitation ends and an attack begins—does not exist:

> *Intent is one of the hardest things to know, but how we define these events matters enormously. The definition will determine the response. If digital events are considered conventional espionage, they may trigger political or legal reviews and approvals. But a digital attack response aimed at disrupting an enemy's capabilities fall more along traditional military lines.*[396]

Military systems are obvious cyber targets. Preventing commanders from communicating with their troops or targeting responses afford attackers major advantages.

Also, critical infrastructure components, while not owned by the military, must also be a central part of national security planning.

Petroleum refineries and nuclear power plants don't need their anti-missile defenses to counter foreign governments, but they do still need cyber defense.

As former Vice Admiral and NSA Director John Mikael McConnell has noted:

> *Information managed by computer networks—which runs our utilities, our transportation, our banking, and communications—can be exploited or attacked in seconds from a remote location overseas. No flotilla of ships or intercontinental missiles or standing armies can defend against such remote attacks located not only well beyond our borders, but beyond physical space, in the digital ether of cyberspace.*[397]

Just as with normal warfare which can begin with limited skirmishes that lead to full-on battles, all cyberattack levels must be taken seriously.

In many cases the computer systems initially targeted may be surveillance operations, malware vulnerability tests, or backdoor battle space preparations for much larger and more devastating assaults.

Writing in *ZDNet Week in Review*, UK *TechRepublic* Editor in Chief Steve Ranger warns that there are endless possible grim scenarios:

> *Perhaps attackers start with the banks: one day your bank balance drops to zero and then suddenly leaps up, showing you've got millions in your account. Then stock prices start going crazy as hackers alter data flowing into the stock exchange.*
>
> *The next day the trains aren't running because the signaling stops working, and you can't drive*

anywhere because the traffic lights are all stuck on red, and the shops in big cities start running out of food. Pretty soon a country could be reduced to gridlock and chaos, even without the doomsday scenarios of hackers disabling power stations or opening dams.[398]

Still, as Ranger points out, there are—thankfully—vanishingly few examples of real-world cyberwarfare, at least for now:

> *Nearly every system we use is underpinned in some way by computers, which means pretty much every aspect of our lives could be vulnerable to cyberwarfare at some point, and some experts warn it's a case of when not if.*
>
> *Governments are increasingly aware that modern societies are so reliant on computer systems to run everything from financial services to transport networks that using hackers armed with viruses or other tools to shut down those systems could be just as effective and damaging as traditional military campaign using troops armed with guns and missiles.*
>
> *Unlike traditional military attacks, a cyberattack can be launched instantaneously from any distance, with little obvious evidence of any build-up, unlike a traditional military operation. Such as attack would be extremely hard to trace back with any certainty to its perpetrators, making retaliation harder.*

As a result, governments and intelligence agencies have strong reasons to worry that increasingly sophisticated hackers will discover ways to bypass traditional or outdated defenses in attacks against

235

vital infrastructure such as banking systems and power grids.

Here, the first line of defense—and offense as well—as to both pursue intelligence gathering while blocking or interfering with those same activities of all adversaries.

Collateral risks of unexpected consequences require that cyber weapons and tools have to be handled—and deployed—with great care. There is also the further risk that thanks to the hyper-connected world we live in that these weapons can spread and also cause much greater chaos than planned, which is what may have happened in the case of the Ukranian NotPetya ransomware attack.

Modern cyber defense planning and programs depend upon cultivating and deploying means to obtain quality intelligence information that cyber warriors can rely on. Serious ambiguities often arise, however, regarding when a particular cyberattack for such intelligence gathering constitutes an actual warfare act.

Some argue cyberwars will never take place; others argue cyberwar is taking place right now. The truth is in a gray area somewhere in the middle.

Such distinctions, while potentially impactful from a strategic perspective, often become obscured by prevailing political policy considerations. Key among these is the transparent reality that virtually all nations do it...including the U.S. and our allies.

According to the *Washington Post*, after revelations about Russian meddling in the run-up to the 2016 U.S. Presidential elections, President Obama authorized the planting of cyber weapons in Russia's infrastructure. As reported:

The implants were developed by the NSA and designed so that they could be triggered remotely as part of retaliatory cyber-strike in the face of Russian aggression, whether an attack on a power grid or interference in a future presidential race.[399]

According to U.S. intelligence chiefs, more than 30 countries are developing offensive cyberattack capabilities, although most of these government hacking programs are shrouded in secrecy.

Steve Ranger points out that just as nations attempt to deter rivals from attacking in conventional weapons, so countries are developing the concept of cyber deterrence to help to prevent digital attacks from occurring in the first place—by making the cost of the attack too high for any potential assailant:

> *There is one key formal definition of cyberwarfare, which is a digital attack that is so serious it can be seen as the equivalent of a physical attack.*
>
> *To reach this threshold, an attack on computer systems would have to lead to significant destruction or disruption, even loss of life. This is the significant threshold because under international law, countries are allowed to use force to defend themselves against an armed attack.*
>
> *It follows then that, if a country were hit by a cyberattack of significant scale, the government is within its rights to strike back using the force of their standard military arsenal: to respond to hacking with missile strikes perhaps.*[400]

So far it's not entirely clear that any cyber-attack has ever reached that threshold. Nor is it certain if and how the "international community" (however we might define that term) would agree upon any specific metric to determine when an official threshold-level "act of cyberwarfare" has been committed.

Presently, there is no legal international framework governing cyberwar other than various scattered, piecemeal agreements that are open to interpretation and exploitation.

A NATO-affiliated group called the Cooperative Cyber Defense Centre of Excellence (CCDCoE) has been working for years to come up with guidelines to bring laws governing traditional warfare into the cyber age. The idea is that by making the law around cyberwarfare clearer, there is less risk of an attack escalating because escalation often occurs when the rules are not clear and leaders overreact.

Aimed at advisers to governments, military, and intelligence agencies, the *Tallinn Manual,* (named after the team's location in the Estonian capital of Tallinn), the document sets out rules regarding when an attack is deemed a violation of international law in cyberspace, and when and how states can respond to such assaults.[401]

Meanwhile, the U.S. government has failed to keep up with threats from abroad.

US intelligence briefings have long warned that Russia has a "highly advanced offensive cyber program" and has "conducted damaging and/or disruptive cyberattacks, including attacks on critical infrastructure networks".

Russia and China have both re-routed Internet traffic to their servers.[402]

Richard Clarke and Robert Knake have reported that the 2019 threat analysis by U.S. intelligence agencies acknowledged that Russia had the ability to disrupt America's power grid and China can take control of the natural gas pipeline.[403]

Clarke and Knake observe that the reason the Russians are in the controls of the U.S. power grid is because it is easy to be, and very useful to be:

> *Whether or not they could actually bring the country to its knees without firing a shot, no one knows. And that is the point.*

They add:

> *We have to admit that it's a hard problem. The power grid is a crazy quilt of hundreds of disparate electric power companies of very different sizes and competencies. Each company has tens of thousands, if not hundreds of thousands, of devices connected to it, many of them sitting out in the open,*

unguarded.[404]

David Sanger offers a dire conclusion in the Afterward section of his book *The Perfect Weapon.* He asserts that although American intelligence officials will not concede the point, internal government assessments say it will be a decade—at least—before the United States can reasonably defend our most critical infrastructure from a devastating cyberattack launched by Russia or China, the two most skilled adversaries in the world.[405]

Arguing that there are simply too many vital networks, growing too quickly, to mount a convincing defense, the offense is still wildly outpacing defense, Sanger quotes cyber expert Bruce Schneider who put the situation this way: We are getting better. But we are getting worse faster.[406]

Marie O'Neill Sciarrone at the George W. Bush Institute agrees that America's military is falling behind in this urgent race for cyber leadership. She writes:

> *The American military does not possess sufficient numbers of skilled operators to counter this growing threat, much less obtain superiority in the cyberspace domain. We will need to train for a different set of skills and knowledge.*[407]

Cyber Geopolitics and Competition

Victor Hvozd, former Chief of the Main Intelligence Directorate of Ukraine's Ministry of Defense and chairman of that country's Foreign Intelligence Service traces key geopolitical cyberspace tensions following soon after the collapse of the USSR.

Writing in *Cyber Conflict and Geopolitics—the Cold War's New Front,* Hvozd, argues that the emergence of new players in the world arena at that time, led by the US, created a multipolar world order dominated by a few centers of power divided among various national interests and lines of communication between them.[408]

This transformative development, in turn, has led to an escalation of confrontation in cyberspace, which is becoming global and can be compared to world wars with the use of weapons of mass destruction.

Hvozd cites evidence of this broadly divisive global instability being witnessed in the spread of massive attacks on computer systems of leading and other countries with critically dangerous consequences in the main spheres of their life (including state management and military command, economy, finance, energy, and transport).

In addition to cyber espionage, Hvozd notes routine use of cyberspace for interference with electoral and political processes and other massive information wars on the Internet. There are real possibilities, he warns, that these cyberwars can and will escalate into trade-economic and sanction wars, as well as into a direct armed confrontation.

Victor Hvozd characterizes a structure of the modern multipolar world system as being divided into four main groups:

The most powerful world-class states, namely, the USA and the PRC., are also the main players in the field of cyber confrontation.

Secondary regional leaders include the EU in the European region, Russia in Eurasia, India in Southeast Asia, Brazil in Latin America, and South Africa in Africa.

Ukraine belongs to a group of nations that primarily realize their interests in cyberspace at local levels and without extensive use of subversive actions.

Pariah states and various extremist and terrorist organizations use cyberspace to carry out subversive activities against other countries and international organizations, including the leading countries.

Hvozd attributes "main lines of confrontation" between different cyberspace nation players that reflect conflicting interests in various political, economic, security and other spheres. According to his world view:

The United States and China compete to maintain ultimate world "domination" [although I wouldn't personally characterize the U.S. goals either as imperialistic or colonialist].

The West and Russia, remain in a de-facto state of "Cold War" as a result of actions by the US, NATO, and the EU to curb Moscow's neo-imperial policies.

The U.S. and their allies are in conflict with attempts of pariah states and extremist and terrorist organizations to gain control over intimidating and destructive cyberwarfare capabilities and other weapons of mass destruction.

China and India are in regional conflict on the basis of the struggle for influence in Southeast Asia.

Israel and Muslim countries are in conflict as a result of the fundamental civilizational controversy and the struggle for Middle East influence, territories, and resources.

Russia and Ukraine and other former USSR's countries of a democratic and European are in conflict with Moscow's attempts to establish control over them.

As a result, Ukraine and its European and Euro-Atlantic allies have become objective allies of the US, NATO and the EU in their confrontation with Russia—including in cyberspace.

The *Worldwide Threat Assessment* of the U.S. Intelligence Community projects that Russia will continue disruptive cyberattacks against the United States and its allies, including Ukraine; China will use cyber-espionage and cyberattacks to support its national security [and, I will add, national citizen control and global thievery]; Iran and North Korea will also create global threats to U.S. interests through the possibility of cyberattacks against the United States.[409]

Former Ukraine Foreign Intelligence Service Chairman Hvozd proposes that the new world geopolitical situation warrants changes in current cyberspace information warfare strategies along lines of principle guided by updated conceptual documents, the National Security Strategy (2017) and the National Defense Strategy (2018) in particular.

Hvozd notes that significant amendments are currently being made to conceptual documents and policies of the leading countries and international organizations regarding confrontation in cyberspace.

The U.S. Department of Defense's Strategy for Operating in Cyberspace (Cyber Strategy) and of the U.S. Department of State's International Cyberspace Policy Strategy, for example, set forth policy directives to:

- Define cyberspace as one of the main environments of the U.S. Department of Defense in providing national security to the United States of America.

- Provide the USA with the right to conduct all types of military operations in cyberspace to defeat any adversary and prevent threats to the country.

- Determine appropriate strategies and tactics for offensive and defensive cyberattacks.

- Outline a list of cyber threats to the USA in the economic, military, social and humanitarian spheres.

- Demonstrate U.S. readiness to use all possible means, including military actions to protect its cyberspace.

- Determine future best-course directions for U.S. cooperation with other countries on joint cyberwarfare planning and actions.[410]

Victor Hvozd notes some examples to illustrate progress regarding ways that leading countries and organizations are improving capabilities and formations of their respective cyberwarfare resources.

- USA: The 2017 elevation of the status of the Cyber Command to the level of a unified combatant command has been tasked to: Plan, develop and conduct intelligence, defense and offensive operations in cyberspace; Secure information networks of the U.S. Department of Defense and the national intelligence community; Assume operational control of forces and means allocated from the branches of the armed forces; and Coordinate the work of specialized cyber security units of the U.S. Department of Defense.

- NATO: New bodies have been introduced, including the NATO Communications and Information Agency, the NATO Cooperative Cyber Defense Centre of Excellence (Tallinn), and a planned NATO Cyber Operations Centre.

- NATO-EU: In 2016 signed a Technical Agreement between the NATO Computer Incident Response Capability Center (NCIRC) and the Computer Emergency Response Team of the European Union (CERT-EU). The agreement provides for joining NATO and EU capabilities for detecting and responding to cyberattacks.

- NATO-Ukraine: In 2016 NATO adopted a Comprehensive Assistance Package for Ukraine which includes protecting the country's critical infrastructure from cyberattacks. Possible forms of Ukraine's participation in the NATO Cyber Defense system may include: Protecting national cyberspace and objects of critical infrastructure from cyberattacks by Russia and all kinds of terrorists and extremists. As noted above, this task is already being fulfilled with NATO's active support; Blocking the use of the Ukrainian information space (servers, communication channels, etc.) by other countries for cyber-attacks against NATO member states. This is important because through Ukraine goes one of the channels of electronic traffic to

243

Europe; Active informational influence on the population of Russia and inhabitants of the occupied territories of Ukraine through e-media. As Ukrainians understand Russia better, it will be more effective than such activity by Western information agencies.

- India: A new military Defense Cyber Agency (DCA) will initially directly employ about 1,000 people drawn from the Indian Air Force, the Army and the Navy, besides from the Integrated Defense Staff (IDS) and will be the precursor to the setting up of a Cyber Command in the near future. The DCA will take various cyber security measures like the Indian Army's indigenously-developed and currently operative Bharat Operating System Solutions (BOSS) system under its wing. BOSS, introduced in 2017, guards the communication and information networks of the Indian Army against espionage. The new body's requirement is being viewed as urgent in the backdrop of the changing modes of warfare which demands cyber expertise and integration. According to experts, future wars will be fought on air, land, sea and importantly in cyberspace and space.[411]

- Russia: An Information Operations Forces within the Ministry of Defense was created in 2013, and a Cyber Command of the General Staff of the RF Armed Forces was established in 2014. The main tasks of these bodies are to centralize the conduct of the cyber information operations; and to protect military computer networks, command and control systems.[412]

- China: Chinese cyber operations, on the other hand, are more scattered across services and ministries. The People's Liberation Army (PLA) General Staff Department is responsible for many cyber espionage operations and manages at least 12 operational bureaus and three research institutes. Two other PLA-authorized forces contain teams of specialists in the Ministry of State Security and the Ministry of Public Security to carry out network warfare operations. Besides, nongovernmental forces that spontaneously engage in network attack and defense can be organized and mobilized for PLA operations as dictated.

Writer Russell Mead reflects in a *Wall Street Journal* column that America's 21st century competition with China is likely to be more dangerous and more complex than occurred in America's old Cold War with the Soviet Union. He attributes this circumstance in part to China's economic power makes it a much more formidable and resourceful opponent than the USSR, and also partly because of dramatic technical changes that have taken place in recent times.[413]

Mead observes that the development of nuclear weapons and intercontinental ballistic missiles which shaped the Cold War resulted in a "balance of terror" that kept the Cold War cold. Neither power was willing to risk total annihilation. Arms-control talks became a centerpiece of superpower relations as both sides sought the nuclear balance.

Today's information revolution has introduced cyber weapons that are much harder to control and more difficult to deter...devastating armaments that can crash their power grids and transportation networks, paralyzing financial systems, and destroy the functionality of anything from hospitals to government offices.

While It isn't hard to know where a nuclear missile comes from, cyberattacks are harder to trace and can easily be pinned on proxies. It is also harder to retaliate—an important key to deterrence.

U.S. companies and government agencies are subjected to daily cyberattacks from a variety of government and criminal groups around the world. It is difficult to determine which attacks warrant retaliatory strikes or other responses, and when appropriate, to calibrate the proper response magnitude. If the retaliation is too weak, it won't deter future attacks. If it is too strong, it may trigger an escalation that could be very hard to control.

Mead cautions U.S. to be wary that deterrence is difficult to establish in the murky, ever-evolving cyber world. Arms control agreements based upon President Ronald Reagan's "Trust but Verify," Cold War policies become far harder to rely upon. He warns:

> *The sheer size of the industrial activities necessary to build up a nuclear force or a missile delivery system made it possible for the U.S. and the Soviet Union to agree on verification measures. Cyberweapon programs are extremely difficult to detect, they bear fruit relatively quickly, and they are cheap compared with nuclear programs and even conventional arms.*
>
> *Even if deals between superpowers can be reached, that won't stop the cyber arms race. Smaller countries, along with well-funded terror and even organized crime groups, can develop significant cyber capabilities.*
>
> *Rapid technological change makes a workable system of cyber arms control even more difficult to achieve. The sophisticated malware of 2012 is junk programming today, and changes in the way the Internet works and how governments and companies use it come so quickly that nobody really knows what offensive and defensive capabilities will be needed in five or 10 years.*[414]

Russell Mead emphasizes that both the government and the corporate sector have big stakes in ensuring and supporting aggressive cybersecurity research and development programs. Here again, there is a major difference between cold War spending which was broadly viewed as a drag on the civilian economy and cyber spending which brings direct benefits to advance a variety of civil tech industries.

Mead reminds U.S. that the Internet itself began as a Pentagon-funded project through the Defense Advanced Research Projects Agency's predecessor and that in the tech world the distinction between the race to achieve supremacy in military useful technology and civilian technology has largely collapsed.

Accordingly, Mead concludes, as large and small powers across the world grasp the degree to which IT is the key to both national defense and national prosperity, defense-related spending must—and will—grow.

Progressively Proactive Presidents

Defense spending naturally follows both what a prevailing presidential administration perceives as a threat level, what they are willing and able to tell the public about what they know, and what they are politically and convincingly prepared to do about mitigating serious threats.

Responsive actions and results of different presidents in concert with their top security advisors reveal very mixed policy positions. Some "behind-the-curtain" insights by informed sources follow.

Ronald Reagan (1981-1989):
As President Reagan announced in a 1982 speech:

> *What I am describing now is a plan and a hope for the long-term—the march of freedom and democracy, which will leave Marxism-Leninism in the ash-heap of history, as it has left other tyrannies which stifle the freedom and muzzle the self-expression of the people.*

247

In pursuit of that plan, "The Great Communicator" may have launched the first incidence of cyberwar against USSR tyranny at a time when the Internet was still in a stage of early infancy.

Richard Clarke and Robert Knake report in their book "Cyber War" that in the early 1980s, the Soviet leadership gave their intelligence agency, the KGB, a shopping list of Western technologies they wanted their spies to steal for them. A KGB agent with access to the list turned it over to the French intelligence service in exchange for a new life in France which was part of the Western intelligence alliance.[415]

Unaware that the exchange had occurred, the KGB kept on working its way down the list.

Once the French gave the list to the CIA, President Reagan gave his okay to "help" the Soviets with their technology needs…but there was a big catch.

Clarke and Knake write:

> *The CIA started a massive program to ensure that the Soviets were able to steal the technologies they wanted, but in doing so introduced a series of minor errors into the designs for such things as stealth fighters and space weapons. Weapons designs, however, were not at the top of the KGB's wish list. What Russia really needed was commercial and industrial technology, particularly for its oil and gas industry.*
>
> *In order to get the product from massive reserves in Siberia to Russia and Western consumers, oil and gas had to be piped over thousands of miles. Russia lacked the technology for the automated pump and valve controls crucial to managing a pipeline thousands of miles long. They tried to buy it from U.S. companies, were refused, and so set their sights on stealing it from a Canadian*

firm.

With the complicity of our northern neighbors, the CIA inserted malicious code into the software of the Canadian firm. When the Russians stole the code and used it to operate their pipeline, it worked just fine initially.

After a while, the new control software started to malfunction. In one segment of the pipeline, the software caused the pump on one end to pump at its maximum rate and the valve at the other end to close. The pressure buildup resulted in the most massive non-nuclear explosion ever recorded, over three kilotons.[416]

In other words, it left an ash pile in history that Reagan had promised.

William Jefferson Clinton (1993-2001):

President Bill Clinton's vision that opening up China's participation in the World Trade Organization (WTO), and sharing of information technology with them would result in a more constructive and militarily benign international partner has turned out very differently than planned.

Instead, assisted by massive theft through cyber espionage of American technologies, those U.S. policies have contributed to the growth of totalitarian Communist Party power and control by police state tactics over all nations of the world.

In March 2000, Clinton predicted that China's membership in the WTO would lead to the elimination of tariffs on information technology products by 2005, making tools of communication cheaper, of better quality, and more widely available.

He also once quipped that:

We know how much the Internet has changed

249

> *America, and we are already an open society.*
> *Imagine how much it could change China. Now*
> *there's no question China has been trying to crack*
> *down on the Internet. Good luck! That's sort of like*
> *trying to nail Jell-O to the wall.*[417]

Twenty years later, China has nailed Jell-O to the wall and is on the verge of producing unprecedented high-technology totalitarianism that not only controls the Internet but may soon corner world markets for advanced technologies of the future, including the revolutionary high-speed 5G telecommunications that will fuel both military power and predatory mercantilism.

As *Washington Times* national security columnist Bill Gertz notes:

> *The prediction that China would be liberalized and*
> *democratized by wiring the country with Internet*
> *access was a huge failure. Today, Chinese repression*
> *is employing American-origin technology to restrict*
> *access to the Internet. And the Chinese are seeking*
> *to expand even tighter controls over the Internet*
> *under the rubric of 'Internet sovereignty.*[418]

Technology transferred to China by the Clinton administration prevented Chinese strategic missiles from blowing up on the launch pad and assisted with technology to launch multiple satellites that now sits atop multi-warhead missiles.

Some of that unregulated U.S. technology transfer to China has instead been incorporated into the PLA's large force of ballistic missiles which are now aimed at American cities and U.S. military bases in Asia.

Bill Gertz reports that the countries involved, Hughes Electronics Corporation and Loral Space and Communications Ltd., simply ended up paying relatively small fines of $32 million and $13 million, respectively, for illegally transferring missile technology that improved the launch reliability of Chinese missiles.[419]

Motorola, under authorization from Clinton administration, reportedly provided China with the technology used in delivering multiple warheads from a single missile, and by 2017 was deploying multiple warheads on its arsenal. According to a previously classified Air Force intelligence assessment, the Motorola satellite launcher could be modified to deploy multiple independently targetable reentry vehicles (MIRVS).

All of China's newest missiles, in fact, are being deployed with multiple warheads.

George Walker Bush (2001-2009):

The Bush administration took a proactive stand to counter China's advancing space weapon superiority.

A January 2001 congressional space commission recommendation that "the U.S. government should vigorously pursue the capabilities called for in the National Space Policy to ensure that the president will have the option to deploy weapons in space to deter threats to, and if necessary, defend against attacks on U.S. interests."

This was followed by U.S. withdrawal in 2002 from a previous US-USSR Anti-Ballistic Missile Treaty (ABMT) which allowed the U.S. to pursue missile defenses against nuclear weapons, including those which are space-based.

In response to these actions, the Chinese established a space defense program, including anti-satellite (ASAT) defense.

In January 2007, China conducted a successful ASAT missile test that utilized a multi-stage kinetic kill vehicle traveling in the opposite direction against a weather satellite target. Several national governments, including the Bush administration, responded negatively to the test highlighting space militarization concerns.

In the initial days after the 2007 ASAT test, China's Communist regime shifted into full denial and deception mode. Bill

Gertz observed that Chinese diplomats around the world at first answered official questions posed by foreign governments about the space test by stating that they had no information—something that likely was not true.[420]

Then in February 2008, the U.S. launched its own ASAT, defending its test mission as necessary to remove threats posed by the decaying orbit of a faulty spy satellite with a full tank of toxic hydrazine fuel.

President George W. Bush also authorized the enormously elaborate joint US-Israeli Olympic Games attack that injected the malicious Stuxnet virus into the computer controllers at the underground Iranian nuclear plant.

Largely the work of the NSA and Israel's Unit 8200, the operation was conceived as a strategy to keep the Israelis focused on crippling the Iranian nuclear program without setting off a regional war.

As *New York Times* national security correspondent David Sanger reported, part of the plan was to slow the Iranians and force them to the bargaining table. "But an equally important motivation was to dissuade Prime Minister Benjamin Netanyahu of Israel from bombing Iran's facilities, a threat he was making every few months."[421]

Sanger notes:

> *Bush took the threat very seriously. Twice before the Israelis had seen threatening nuclear projects under wat, one in Iraq, the other in Syria. They had destroyed them both.*

Also, according to Sanger, Israel's spymaster Meir Dagan who was ousted by Netanyahu had reportedly objected strongly to an air attack on the nuclear plant. His reasoning was that an airstrike would provide only an illusory solution because Iran's program was too sprawling and too deep underground for success. Within a few months, the facilities would be rebuilt.[422]

President Bush encouraged U.S. security forces under future administrations to carry out the Olympic Games cyber campaign against Iran's Natanz nuclear facility.

However, the Bush administration was less willing to destroy Saddam Hussein's financial assets by cracking into the networks of banks in Iraq and other countries. Although Richard Clarke and Robert Knake report that the capability to do so existed, government lawyers feared that raiding bank accounts would be seen by other nations as a violation of international law, and viewed as a precedent.[423]

Barack Obama (2009-2016):

President Obama authorized U.S. NSA, in collaboration with Israeli security forces, to carry out the Olympic Games cyberattack on Iran's Natanz plant. As one of the architects of the plan told David Sanger, "That was our holy grail." [424]

Sanger has also reported that there had been no consensus within the Obama administration about how such weapons should be used. Even while Obama was approving new strikes on the Iranian nuclear plant, he harbored his doubts. He had repeatedly questioned whether the United States was setting a precedent—using a cyberweapon to cripple a nuclear facility—that the country would one day regret.

In his book, *The Perfect Weapon*, Sanger writes:

> *Curiously, Obama had already proven willing to engage in a public argument over similar questions about drones. Everything about drone warfare had been secret when he came to office, but over time Obama made elements of the program public and proved willing to explain the law and reasoning behind his decision to deploy these remote-controlled killing machines.*

> *In doing so, he gradually lifted the security*
> *surrounding the use of drones so the world could*
> *understand whether they were hitting terrorists,*
> *and when they went awry and killed children or*
> *wedding guests.*[425]

In 2013, following three ASAT orbiting killer vehicle demonstrations, the Pentagon was strictly banned by liberal, anti-defense policymakers from publicly condemning the test. Behind-the-scene officials wanted to avoid giving the military justification for building American space arms to challenge and deter the new threat from China in space.[426]

Downplaying the significance, an Obama administration spokesman would only say that all three Chinese spacecraft was being monitored by the Strategic Command's Joint Functional Commandant for Space, "consistent with its routine operations to maintain track of objects in space."

The Obama administration also didn't want to appear to be picking a fight political with Chinese leadership, either in physical space or in cyberspace.

David Sanger observes that in 2013, the frustrated president had been prepared to sign a new executive order to bolster America's response to cyber intrusions, yet he couldn't quite bring himself to name the Beijing government as the chief offender.

> *We know hackers steal people's identities and*
> *infiltrate private emails. We know foreign countries*
> *and companies swipe our corporate secrets. Now*
> *our enemies are also seeking the ability to sabotage*
> *our power grid, our financial institutions, our air*
> *traffic control systems. We cannot look back years*
> *from now and wonder why we did nothing in the*
> *face of real threats to our security and our*
> *economy.*[427]

Sanger observed that every time the Obama White House considered calling the Chinese out for their data hacking thefts, there was a temptation to pull their punches:

> *There was always countervailing interests: The State Department needed help on North Korea, the Treasury didn't want to upset the bond markets, the markets didn't want to see a trade war started.*
>
> *In the cyber realm, this meant holding back on naming the Chinese when they got caught in some of the biggest hacks in recent years.*
>
> *Instead, objections would be raised with Chinese in closed sessions at the annual 'Strategic and Economic Dialogue,' assuring that any discussion would remain quiet. And they would almost always result in a scripted Chinese response: It's not us, the officials would insist. It's a bunch of teenagers, or criminals, or miscreants.*[428]

Bill Gertz, in his book *Deceiving the Sky*, agrees with Sanger that appeasement of America's adversaries became the keystone Obama administration foreign policy priority:

> *Under Obama, American security was damaged by ignoring and then covering up China's massive theft of American technology.*
>
> *Abroad, inaction by Obama facilitated the Chinese domination over the strategic South China Sea. After reclaiming 3,200 acres of new islands, China deployed advanced anti-ship and anti-aircraft missiles on the island in what the CIA described as China's Crimea, a stealth takeover similar to the Russian annexation of Ukraine's peninsula in*

2014.[429]

Gertz observes that by 2016 the Obama White House went so far in appeasing Beijing as to prohibit all officials from publicly talking about the military and other threats from China.

Meanwhile, although according to Richard Clarke and Robert Knake the USSR's December 2015 cyberattack on Ukraine had "scared the hell out of American defense officials," they were determined not to show it.

Even in their most benign mode, the implants are useful for surveillance—broadcasting back to their home base news about what is happening inside a network. Those same implants, however, can also be repurposed as weapons simply by injecting new codes.

> *So on one day, the implant may be sending back blueprints on the electric grid. The next day it can be used to fry the grid. Or wipe out data. Or allow someone in a remote locale to take control of the equipment—and drive it off a cliff, so to speak.*[430]

David Sanger concurs that no one in the Obama White House could possibly know whether the Russians intended the Ukraine hack to be a nuisance, an attack, a warning, or a rehearsal for something much larger:

> *Perhaps they were just exploring how easy—or difficult—it was to get inside America's electric utilities, each of which is configured differently.*[431]

After all, officials reasoned, the Russians—and other nations—had been lurking inside American utilities, financial markets, and cell phone networks for years. But at least so far (cross your fingers) they hadn't hit the kill switch.

In any case, the Obama administration almost reflexively

decided to treat the early breaches in the American utilities' networks as a classified secret:

> *Senior members of Congress, selected staffers, and utility company CEOs were taken into locked, signal-proof rooms for briefings on the intelligence. There was no note-taking.*

Under strict rules they were given, the utility executives were barred from sharing the information with the people who administered their networks. Put another way, the only people who could do something about the problem—or at least prepare backup systems—were prohibited from knowing about it.

According to the *New York Times* national security correspondent Sanger, the intelligence agencies said they feared that if the discovery were made public it would tip off the Russians to the quality of our detection systems and perhaps how deeply the NSA is into Russian systems as well.

Sanger acknowledges that this was indeed a risk, but then asks:

> *But could anyone imagine the United States withholding similar intelligence about an impending terror attack that might bring down a bridge, or blow up an electric substation?*
>
> *Almost certainly not.*
>
> *For the Russians, there were no consequences in Ukraine or the United States for hacking into the grid.*[432]

This situation would play out time and again: The NSA did not want to expose intrusions in American systems, for fear of exposing 'sources and methods.' The White House did not want to reveal what was known for fear that, as one of Obama's top advisors put it, 'someone will then ask, what do you plan to do about it?"

China and Russia weren't America's only cyber and ballistic threats.

By mid-2013, it had become clear to the Obama administration that it was simply too late to stop North Korea's bomb production.

While the Iranians were still struggling to make centrifuges spin to produce uranium, the North Koreans were churning out atomic bombs. Although intelligence estimates differed, the North Koreans already possessed upward of a dozen nuclear weapons—and production was speeding up.

Early in 2014, Obama, the Pentagon and various intelligence agencies decided to step up a series of cyber-and electronic-warfare strikes on Kim's missiles, starting with an intermediate-range missile called the Musudan. The plan was either to sabotage them before they got into the air or to send them off-course moments after launch.

It was further hoped that the North Koreans, like the Iranians before them, would blame themselves for manufacturing errors.

Obama reportedly warned that it would take a year or two before anyone would know if the accelerated program could work. And at the same time, Kim had a cyberattack plan of his own.

Barack Obama's target was North Korea's missiles; Kim's was the studio in London where they were producing a television series featuring an American president and a British prime minister who joined forces to free a nuclear scientist kidnapped in Pyongyang.

Within weeks, things started going wrong when became clear that someone had hacked into the TV channel's computer systems. Only years later did anyone notice the similarities between this and what was happening more than five thousand miles away in a hack on Sony.

Once the North Koreans hackers had burrowed into the Sony networks in the fall of 2014, they littered the company with phishing emails, betting that someone in the studio would click on the bait. It didn't take long for the tactic to pay off.

The Sony hack was a harbinger—a destructive attack that

melted physical equipment utilizing only ones and zeros, as Stuxnet had done; a distracting release of private communications that dominated the news and upended careers; and a ransom demand that distracted from the real purpose of the operation, to intimidate and ridicule the Obama administration.

A North Korean hacker group logged in on Facebook *as* Guardians of Peace issued a declaration that the movie's premiere in New York could be the target of a terror attack. It warned:

> *Soon all the world will see what an awful movie Sony Pictures Entertainment has made. The world will be full of fear, Remember the 11th of September 2001.*

The scare tactic worked. Under pressure, the movie release was suspended by Sony.

As Sanger notes, ever cautious President Obama came to the conclusion that Sony wasn't really terrorism. No, it was more like "cyber vandalism." A few days later he admitted not wanting to escalate the situation.

Kim had correctly calculated that a cyber-arsenal was the great leveler. It was dirt cheap, he could launch it from outside the country, and unlike his nuclear arsenal, cyber weapons could be untraceably used to embarrass against his greatest enemy— America—without triggering catastrophic retaliation.

David Sanger conjectures:

> *And even if the United States was willing to retaliate, Kim calculate, doing so would not be easy. To most of the world, the absence of computer networks, of a wired society, is a sign of backwardness and weakness. But to Kim, this absence created a home-field advantage.*
>
> *A country cut off from the world, with few*

> *computer networks, is a lousy target; there are simply not enough 'attack surfaces,' the entry points for inserting malicious code, to make a retaliatory cyberattack on North Korea viable.*

How do you turn out the lights in a country that doesn't have enough power to turn them on?

The NSA was already well aware of North Korea's active hacking operations at least four years before the Sony hack.

As revealed in Snowden WikiLeaks documents. a secret U.S. program code-named "Nighttrain" had opened a cyber-surveillance window into North Korea's intelligence networks. Without telling our South Korean allies, the agency had successfully drilled into the networks that connect North Korea to the outside world, mostly through China and tracked down North Korean hackers, some of whom worked from Malaysia and Thailand.

"The Perfect Weapon", David E. Sanger, 2018, Crown Publishing Group, Penguin Random House LLC, New York

In his book *"The Perfect Weapon"*, David Sanger, who teaches national security policy at Harvard's Kennedy School of Government, discusses a political conundrum that Kim's Sony hack dilemma posed for Obama and his top intelligence advisors.

Obama, meeting in the Situation Room the day the Guardians of Peace threat was delivered, realized he could no longer remain silent. If he ignored a crude threat of terrorist action against theaters, he would look weak.

On the other hand, Obama also recognized pressure to call out the North Korean leadership, blame them for the cyberattack, and make clear what would happen is theaters were attacked. Doing so, however, would publicly reveal a top-secret that the attack and threats had been linked to Kim Jong-un.

Understandably, Obama's intelligence officials were adamant that he could not reveal the presence of any implants the United States or South Korea had lurking in North Korea's systems. Nor did they want to put him in a position of having to explain how they had matched the hacking tools used against Sony to others previously traced—at least tentatively—to North Korea's Bureau 121.

On December 19, 2014, Obama did take at least a baby-step forward step in blaming North Korea for the attack and vowing that a proportional response would happen "in a place and time and manner that we choose."

Leaving no doubt that he was directly challenging Kim Jong-un, Obama said, "We cannot have a society in which some dictator someplace can start imposing censorship here in the US."

He also took a shot at Sony for pulling the film out of theaters. American filmmakers and distributors, he said, should not "get into a pattern where you're intimidated by these kinds of criminal attacks."

Obama administration retaliation to the Sony cyberattack was anemic at most. As Sanger observes:

> *In early January, the White House announced some weak economic sanctions against North Korea—sanctions which Kim may never have noticed, given how many others had already been imposed. But afterward, the official who designed that 'proportional response' acknowledged that it was ridiculously weak, given the gravity of the attack on Sony and the president's vow that the United States would not tolerate intimidation.*

Economic sanctions, the first tool of presidents who want to show they are doing something without risking conflict, might have been satisfying on the day they were announced. But there was no evidence that sixty years of sanctions had slowed, much less stopped, the North Korean's assembly of a good-sized nuclear stockpile.

Why would sanctions do any better against a cyberattack?

Sanger observes that Obama's unwillingness to confront foreign cyber threats against U.S. corporations constituted a repeated national security weakness.

Obama and his aides had previously concluded that the denial-of-service attacks didn't rise to the level of requiring a national response when Iranian hackers froze the banking networks of Bank of America and JPMorgan Chase in 2011 and 2012. The attacks were instead viewed as crimes, not terrorism, and referred to the Justice Department, which ultimately indicted Iranian hackers.

The Obama administration also chose not to respond in early 2014 when Iranian hackers melted down computer equipment at the Sands Casino in Las Vegas. The reported attack motive was to show owner Sheldon Adelson that if he wanted to advocate setting off a nuclear weapon in the Iranian desert, he had better be prepared to see his prize casino go offline.

That cyber assault was once again treated by the White House as merely a criminal act—something to be dealt with in the courts—rather than as an attack on the United States.

David Sanger concludes that during eight years in the White House—years in which cyber went from a nuisance to a mortal threat—neither Obama nor the U.S. national intelligence bureaucracy ever formulated a comprehensive national security policy response.

The Sony attack, for example, had focused the administration's attention on North Korea, but on its cyberattacks, not its missile program. Similar policy confusion occurred as Iran's cyberattacks presented inconvenient policy controversy at a time when the Obama State Department was conducting negotiations that ultimately led to the 2015 "nuclear deal" which also excluded missile delivery systems.

Obama seemed fatalistic about Russia's cyberattacks on Ukraine in 2015 and 2017. On February 15, 2017, he told Jeffrey Goldberg of *The Atlantic*, "The fact is that Ukraine, which is a non-NATO country, is going to be vulnerable to military domination by Russia no matter what we do." [433]

The president was similarly cautious about Syria when the

Pentagon and NSA came to him late in the spring of 2011 with a battle plan that featured a sophisticated cyberattack on the Syrian military and president Bashar al-Assad's command structure.

For Obama, who had been adamantly opposed to direct American intervention in a worsening crisis in Syria, such methods would seem to be obvious, low-cost, low-casualty alternative. Nevertheless, he declined the idea, determining that Syria was not a place where he saw strategic value in American intervention—not even covert attacks.[434]

The Obama administration had been engaged in a largely secret debate about whether cyber arms should be used like ordinary weapons, whether they should be rarely used covert tools, or whether they ought to be reserved for extraordinarily rare use against the most sophisticated, hard-to-reach targets.

Looming over the issue was the question of retaliation: whether such an attack on Syria's airpower, its electric grid or its leadership would prompt Syrian, Iranian or Russian retaliation in the United States.

The Obama White House also passed upon an opportunity to take firm action against Beijing after a series of intellectual property cyber thefts—with China as lead suspect—on companies such as Mitsubishi, AT&T, Visa, and Nissan. Keith Alexander, the NSA's director at the time, referred to the events as "the greatest transfer of wealth in history."

Time and again Obama's "lead from behind" military and cyber strategies have clearly failed. Although he and President Xi signed an agreement in 2015 to limit cyberattacks on each other for commercial purposes, Chinese penetrations of U.S. organizations continue.

Donald J. Trump (2016-2024):
U.S. presidential cyber-security policies that had remained slanted in favor of appeasing China and other adversarial nations ended abruptly with the election of President Donald Trump.

In 2018, clarifying that Beijing remains to be America's main cyber nemesis, the Trump administration, for the first time, publicly exposed activities of one of China's most important spymasters, General Liu Xiaobei.

Li's key 3PLA unit within the Communist Party Cyber Corps was the primary organization behind a relentless campaign to steal some crown jewels of American economic and defense secrets.

As discerned by *Washington Times* national security columnist Bill Gertz, failures of successive presidential administrations to publicly tell the truth about a growing Communist Chinese menace were based upon fraudulent theories and analyses that left the United States and its allies around the world in precarious position of trying to play technological catch-up.[435]

Gertz writes:

> *Until 2016 and the advance of the Donald Trump presidency, details about Chinese cyberattacks and the organizations behind them were tightly held secrets. Successive administrations since the 1990s sought to cover up and hide nefarious Chinese intelligence activities as part of rigid policies to appease Beijing.*
>
> *It was through such feckless, defeatist policies that the United States theorized that conciliation and engagement would lead the Party rulers and their military henchmen away from communism and toward democracy and free markets. Instead, a hated Communist Party regime was not only perpetuated but strengthened at the expense of America's most precious intellectual resources.*[436]

A no-nonsense 2017 National Security Policy document issued by the Trump administration threw down the gauntlet with Beijing by declaring protection of satellites a vital interest that the U.S. was willing to go to war to defend. The strategy stated:

Any harmful interference with or attack upon critical components of our space architecture that directly affects this vital interest will be met with a deliberate response at a time, place, manner and domain of our choosing.[437]

Just over a decade before, in 2007, cyberattacks were missing entirely from the global "Threat Assessment" that intelligence agencies prepared each year for Congress. Instead, terrorism and psst-911 concerns had understandably topped that list.[438]

That priority hierarchy had changed as a result of a variety of cyber threats ranging from paralyzing ransomware attacks on American cities to DDoS assaults on corporate organizations and government institutions.

On September 20th 2018, the Trump White House National Security Council issued the first National Cybersecurity Strategy policy in 15 years. Articulated priorities are to secure Federal networks and information, secure critical infrastructure, combat cybercrime, and improve incident reporting.

A year into Trump's presidency, his defense secretary, four-star General Jim Mattis, had issued a warning that nations around the world were threatening to use cyberweapons to bring down America's power grids, cell phone networks, and water supplies.

Mattis urged the president to declare that he was ready to take extraordinary steps to protect the country. All nations should be forewarned that any attack on America's critical infrastructure, might be met with a nuclear response.

General Mattis, whose career featured two decades in the Middle East chasing Al Qaeda and ISIS starkly stated in early 2018 that America's "competitive edge has eroded in every domain of warfare," including the newest one, "cyberspace."

Rob Joyce, Trump's cyber czar for the first fifteen months of the administration who previously spent years running the NSA's Tailored Access Operations unit, had previously recognized ever-increasing U.S. cyber threats.

He said in 2017:

> *So much of the fabric of our society rests on the*
> *bedrock of our IT. We continue to digitize things;*
> *we store our wealth and treasure there; we run our*
> *operations; we keep our secrets all in the cyber*
> *domain.*

In short, we are inventing new vulnerabilities faster than we are eliminating old ones.[439]

China wasn't the only major adversarial opponent facing a serious new White House cyber sheriff. The Trump administration's original CIA director, Mike Pompeo, had occasionally hinted at "working diligently" through ongoing programs to slow Kim's progress and delay the day when he was ready to put a nuclear warhead atop one of his missiles. Pompeo sometimes suggested that day might otherwise be just "months away."

American intelligence officials were caught off guard at how rapidly Kim's regime had rolled out an entirely new long-range missile technology that could reach Guam, then the West Coast, then Chicago and Washington, DC.

In September 2017, North Korea detonated a sixth nuclear bomb, one that was fifteen times more powerful than that leveled Hiroshima.

Trump almost immediately requested $4 billion in emergency congressional funding to boost missile defense and other "disruption/defeat" efforts to contain North Korea threats. Budget priorities included sophisticated cyber and electronic strikes.[440]

Meanwhile, it was also already well known that the Russians had targeted American nuclear power plants as well as water and electric systems, with the insertion of malicious code that would give Russia the opportunity to sabotage the plants or shut them down at will.

Yet as David Sanger observes:

Accustomed to an era when battles were fought by soldiers, the international community was hesitant to act without ascertaining the same level of attribution that insignia once provided.

His green men and hackers alike were cloaked in enough ambiguity that Putin could get away with attacks consequence-free—even when there was little doubt that he was the source.[441]

Wars of the future will be won or lost by armies with the smartest rather than simply the most powerful weaponry. Outpacing particularly dangerous adversaries such as China, Russia, North Korea, and Iran will require rapid advancements in AI and quantum science, including quantum navigation to relieve militaries of reliance on GPS satellites and space.

A National Quantum Initiative Act enacted in 2018 under Trump leadership has committed $1.25 billion over five years to advance U.S. competitiveness in this area. By sad comparison, Beijing, America's chief rival, spends at least $2.5 billion per year on quantum research—more than 10 times what Washington spends.[442]

China has vowed to draw equal with the U.S. in AI within two years, overtake it by 2025, and become dominant in the world force by 2030. They have established a massive quantum center in Hefei province, and aspire to develop a cyber-code-breaking 'killer app" which will be capable of rendering most current public-key encryption architectures—used for applications from banking and credit cards to the power grid—obsolete.

As previously mentioned, China launched the world's first quantum satellite in 2016. They also created a quantum-encrypted intercontinental video link from space to a China-Austria study group in Vienna.

Besides, China created a 1,263-mile ground link between Beijing and Shanghai using quantum-encrypted keys between relay stations, which offers an ultra-secure network for transmitting sensitive data, including China's military and intelligence services.[443]

Meanwhile, Google has reported that its quantum computer, dubbed "Sycamore", solved a mathematical calculation in 200 seconds that would take a supercomputer 10,000 years.

IBM then asserted that its "Summit OLCF-4" at the Oak Ridge National Laboratory was hot on Google's trail and could have done the same calculation in 2.5 days—roughly a thousand-fold rather than 1.5 trillion-fold difference.[444]

Given the top-secret, proprietary and revolutionary nature of quantum developments in America and abroad, any ultimate victors of this cyber-supremacy competition will become known only by calamitous surprise.

That's an enormous victory wager that America can't afford to lose.

Assessing Risk and Responses

As pointed out by Martin C. Libicki in a 2009 Rand report *Cyberdeterrence and Cyberwar* prepared for the U.S. Air Force, cyber war and conventional war capabilities are fundamentally different:

> *Cyberspace is its own medium with its own rules.*
> *Cyberattacks, for instance, are enabled not through*
> *the generation of force but by the exploitation of*
> *the enemy's vulnerabilities.*

Cyberwar strategies require that both offensive and defensive capabilities remain held as secret surprises. Each side works to prevent the other from knowing what it knows about them, their capabilities, and their vulnerabilities.

Major challenges exist in assessing whether a nation they are targeting will surprise them with a significantly improved array of defenses in a crisis...like disconnection of its weapons from international cyberspace. On the other hand, they also can't always be certain if trapdoors and logic bombs placed in an adversary network have been discovered, neutralized, and secretly traced back to spy on them.

If the cyber warrior's job, for example, is to shut down an enemy's air defense system slightly in advance of his nation's air force doing a bombing mission, failure can lead to a reverse surprise. The radar installations and missiles that were supposed to have been shut down may suddenly come alive and destroy the attacking aircraft.

Risk and retaliation correlations are further complicated by uncertainties regarding an adversary's motives and ultimate goals. As previously discussed, most government and corporate cyber incidents thus far have involved network penetrations to steal information or to implant trapdoor surveillance bugs. In many cases, they occurred unnoticed or intentionally unreported.

Even subversive purposes of DDoS and ransomware attacks are often difficult to ascertain...whether to attribute to rogue criminals versus nation-state actors.

Richard Clarke and Robert Knake point out in their book *"Cyber War"* that detecting whether and why a nation is engaging in cyber espionage may often be close to impossible. Then, even when high-confidence verification of a violation and attribution can be achieved, the cover purpose may still have been intentionally misleading to hide other activities.[445]

Clarke and Knake argue that the threat of retaliation for moderate-level cyberattacks—presumably short of major damage to critical infrastructure and other safety-critical systems—has limited deterrence impact:

> *In the case of cyber war, the power of the offense is largely secret; defenses of some efficacy could possibly be created and might even appear suddenly in a crisis, but it is unlikely any nation is effectively deterred today from using its own cyber weapons*

*in a crisis, and; the potential of retaliation with
cyber weapons probably does not deter any nation
from pursuing whatever policy it has in mind.*

*So, we cannot deter other nations with our
cyber weapons. In fact, other nations are so
undeterred that they are regularly hacking our
networks. Nor are we likely to be deterred from
doing things that might provoke others into making
a major cyberattack.*

*Deterrence is only a potential, something that
we might create in the mind of possible cyber
warriors if (and it is a huge if) we got serious about
deploying effective defenses for some key networks.
Since we have not even started to do that,
deterrence theory, the sine qua non of strategic
nuclear war prevention, plays no significant role in
stopping cyberwar today.*[446]

And whereas in the physical world, the territory is relatively fixed,
the amount of "ground" that the defender has to protect is
constantly growing, and growing exponentially, in the cyber world.

The first step in "preparing the battlefield" for attack invariably
involves infiltrating networks, gathering information, and potentially
even laying the groundwork for more aggressive action…just as
occurred with Operation Orchard, the 2007 Israeli strike on the
nuclear research facility in Kibar, Syria.

These battlefields are now truly global.

As Israeli Chief of Military Intelligence Major General Amos
Yaldin, observed: "Cyberspace grants small countries and individuals
a power that was heretofore the preserve of great states."

Speaking at a December 15th 2009, Institute for National
Security Studies (INSS) conference in Tel Aviv, General Yaldin listed
vulnerability to hacking among Israel's greatest national threats.
Citing the Iranian nuclear program, and Syrian and Islamic guerrillas

along Jewish borders as examples, he explained:

> *The potential exists here for applying force ... capable of compromising the military controls and the economic functions of countries, without the limitations of range and location.*[447]

And as Mike McConnell who served as director of U.S. national intelligence from 2007 to 2009 testified to the U.S. Senate, "if the nation went to war today, in cyberspace, we would lose. We're the most vulnerable. We're the most connected. We have the most to lose."

McConnell explains that unlike many of its potential adversaries in this space, the United States, it's military, in particular, is highly reliant on computer networks for everything from communications to electricity. The vast majority of electrical power used by U.S. military bases, for instance, comes from commercial utilities using the same power grid.

Accordingly, cyberattacks of equivalent strength would have far more devastating consequences on the U.S. than on potential adversaries like China, whose military is still less networked, let alone North Korea, whose economy never even entered the information age.

Former NSA official Charlie Miller agrees:

> *One of North Korea's biggest advantages is that it has hardly any Internet-connected infrastructure to target. On the other hand, the United States has tons of vulnerabilities a country like North Korea could exploit.*[448]

US military computers are believed to be regularly targeted by Chinese PLA hackers seeking such information as unit deployment schedules, resupply rates, material movement schedules, readiness assessments, maritime prepositioning plans, air tasking for aerial refueling, and the logistics status of American bases in the Western Pacific Theater.

And as the 2013 Snowden leaks showed, U.S. intelligence operations harvest similar information about potential adversaries in China and elsewhere.[449]

Once inside military communications networks, a foe can disrupt or even disable command and control, keeping commanders from ending out orders, units from talking to each other, or even individual weapons systems from sharing needed information.

For example, more than one hundred American defense systems, from aircraft carriers to individual missiles, rely on the Global Positioning System (GPS) to locate themselves during operations.

An accidental 2010 software glitch knocked 10,000 military GPS receivers offline for over two weeks, including the U.S. Navy's X-47 prototype robotic fighter jet. As Singer and Friedman warn, cyberwarfare, in effect, makes such software glitches deliberate.[450]

Vastly increasing and diverse robotic surveillance devices and weaponry add both to offensive capabilities and the number of computer network cyber targets. The US, for example, operates many thousands of remotely-controlled military "drones" such as the famous "Predator" and "Reaper". More than eighty countries reportedly have military robotics programs.

The same computer networks and GPS systems that enable drones to strike targets thousands of miles away also open up endless possibilities to remotely disrupt or co-opt those interconnected devices.

Alternatively, rather than attempt to disable or jam opponent communications, an attack may target information within them, feeding the perpetrator false reports via its own devices. The goal of this "information warfare" is to get inside the enemy's mind to influence decision-making.

Strategic objectives of such operations can range from issuing false military unit commands from top leaders to more tactical insertions aimed at compromising individual weapons systems and their sensors as the Israelis did in Operation Orchard.

Compromised robotic systems can be "persuaded" to do the opposite of what their owners intend. This creates a whole new type of combat, where for example, a goal may not be merely to destroy the enemy's tanks but to hack into his computer networks and make his tanks drive around in circles or even attack each other.[451]

Such capabilities are no longer merely theoretical. Many military systems operate under Supervisory Control and Data Acquisition (SCADA) programs that can be targeted in much the same way that the US-Israeli Stuxnet attack caused Iranian centrifuges to spin out of control.

Richard Clarke and Robert Knake point out that the clearest example of the dependency and vulnerability brought on by SCADA software and transmission controls also happens to be one system that everything depends upon the electric power grid.[452]

SCADA software programs control electrical loads used by various devices which include those that regulate critical power grids. These signals are often sent via internal computer networks and sometimes by wireless and radio access.

Many of the electric grid devices also have multiple connections. Nearly half are directly linked to the Internet... meaning that someone can remotely hack into them from anywhere on the planet.[453]

Clarke and Knake trace back this dangerous vulnerability to U.S. deregulation legislation in the 1990s when electric power companies were divided up into generating firms and transmission companies which were allowed to buy and sell power to each other anywhere within one of the three big grids in North America.

At the same time, they were, like most every other computer-networked company, they installed common commercially-available software programs to manage their purchasing, sales, power generation, and transmission. SCADA systems, which were already running each electric company's substations, transformers, and generators at the time, became interconnected with everything else...including remote diagnostics.

Making grid vulnerabilities potentially even more impactful, the enormous generators that power the United States are manufactured when they are ordered, on the just-in-time delivery principle. This means that they aren't just sitting around to be sold. A big generator that becomes badly damaged or destroyed is unlikely to be replaced for months.[454]

A 2008 "Smart Grid Initiative" enacted by the Federal Electric Regulatory Agency attempted to reduce grid vulnerability by requiring electric companies to adopt specific cyber security measures or risk noncompliance fines. Clarke and Knake note that an unfortunate regulation down-side is to cause the electric grid to become even more wired... even more dependent upon computer network technology.

Peter Singer and Alan Friedman note that just as the growing number of Internet-connected computers introduces hacking vulnerabilities, so does the phenomenally rapid expansion of coding complexity now constituting many millions of item software lines.

By comparison, malware has stayed relatively short and simple (some as succinct as just 125 lines of code), and the attacker only has to get in through one node at just one time to potentially compromise all defensive efforts. This creates a circumstance where defensive efforts require far greater investments than those of attackers.[455]

In their most recent book, *The Fifth Domain*, Clarke and Knake opine that commercial companies must take on more of those defensive responsibilities and costs. They write:

> *Since the Clinton administration, our cyber strategy has changed very little despite many attempts to come up with a different one.*
>
> *Thus, when we consider how to secure*

*privately owned and operated networks, we return
to the basic idea that the companies that own and
operate the Internet and the things connected to it,
be they multinational media companies, providers
of essential services, or the makers of the tiniest IoT
devices, will be responsible for protecting
themselves. They will do it through network
defense, not offense.*[456]

Clarke, who served as White House security staff for George H.W.
Bush, Bill Clinton, George W. Bush; and Knake who served on
Obama's staff, observe:

*Over the last thirty years, the U.S. government has
worked to get out of the business of running the
Internet, turning the operation over to the
backbone to commercial providers in 1995.*

*The final piece of transitioning the operation of
the Internet to the private sector took place in the
fall of 2015, when the Commerce Department
ended its contract for the operation of the Internet
root servers, the systems that allow the translation
of domain names like goodharbor.net to the 1s and
0s that computers understand.*

Clarke and Knake concludes:

*As a nation, we should be hesitant about inviting
the government back onto the network.*[457]

Perhaps the same good advice might apply to keep government out
of our personal lives.

When Government Invades Our Privacy

My most recent book *The Weaponization of AI and the Internet* urges readers to consider the fate of a frog in a shallow pan of water placed over a flame complacently adjusting to the temperature change until it's too late to jump out.

Most particularly, the book implores U.S. all to be very cautious of all-too-seductive invitations to trade away precious privacy and liberty for promises of increased convenience, efficiency, and security from predators.[458]

If Edward Snowden's 2013 release of a massive trove of WikiLeaks documents revealing the CIA and UK's MI5 maintained an armory of surveillance tools to spy on private citizen's smart TVs, cars and cellphones weren't bad enough, then-FBI Director James Comey's declaration supporting such activities fanned the flames of public outrage.

Speaking at a Boston College conference on cybersecurity, Comey said:

> *There is no such thing as absolute privacy in America: there is no place outside of judicial reach.*[459]

WikiLeaks files provided clear evidence that the CIA had assembled a formidable arsenal of malware designed to target operating systems of mobile phones, laptops, and TVs including Android, iOS and Windows. The documents also showed broad exchanges of tools and information between the CIA, the National Security Agency, and other U.S. federal agencies, as well as intelligence services of close allies Australia, Canada, New Zealand and the United Kingdom.

One document dealing with Samsung televisions carries a CIA logo and is described as a secret. It adds "USA/UK" and says "Accomplishments during joint workshop with MI5/BTSS (British Security Service). It details how to fake it so that the television appears to be off but in reality can be used to monitor targets. It also describes the television as being in the "Fake Off" mode.

In other words, the "Fake Off" mode leads the owner to falsely believe that the TV is off when it is on, operating as a bug that records room conversations that are sent over the Internet to a covert CIA server.

Referring to concealed UK involvement, the document says: "Received sanitized source code from UK with comms and encryption removed."

"We can't spy on our own citizens but we can spy on anyone else's," explained Neil Richards, a law professor from Washington University. "If agencies are friends with each other, they have everybody else do their work for them and they just share the data."

After all, the idea wasn't new at all. The NSA had developed a "clipper chip" in the 1990s that could be secretly installed in computers, TVs and early cell phones to encrypt encrypted voice and data messages with a back door that the agency could unlock. It enabled intelligence agencies with proper authorization including the FBI and local police to decode any message.

The Clinton administration endorsed the idea—at least for a while—arguing that once the chip went into every device, there would be no way for terrorists to avoid being monitored.

The plan was scrapped after most manufacturers rebelled.[460]

Following angry public reactions to the Snowden revelations, President Obama created an independent panel to advise his administration regarding what kind of restrictions—if any—to put on NSA in balancing public privacy and security concerns. The panel made a unanimous recommendation that the government should "not in any way subvert, undermine, weaken, or make vulnerable generally available software."

Instead, the government should simply "increase the use of encryption and urge U.S. companies to do so."

NSA immediately urged Obama to ignore the panel's advice, warning that allowing personal encryption would create a "going dark" crisis that would keep its agents—along with local police—from tracking terrorists, kidnappers and spies.

Further, as David Sanger notes, the FBI argued that authorization of a court-approved back door into Apple's phones was also essential to thwart ISIS plotters and homegrown terrorists.

Apple's CEO Tim Cook believed that granting the U.S. government a back door demanded in its products would be a disaster. He told Sanger:

> *The most intimate details of your life are on this iPhone. Your medical records. Your messages to your spouse. Your whereabouts every hour of the day. That information is yours. And my job is to make sure it remains yours.*[461]

The burden of arguing for government access to personal information fell on then- FBI Director Comey who cited the most emotive example he could come up with to support his position.

Comey asked what would happen when the parents of a kidnapped child came to him with a phone that might reveal the whereabouts of their child, but its contents could not be determined because they were automatically encrypted—just so that Apple could extend its profits around the world?

The FBI director then predicted that there would be a moment, soon, when those parents would come to him "with tears in their eyes, look at me, and say, 'What do you mean you can't'" decode the phone's contents?

Comey continued, offering an analogy to children trapped in apartment doors and car trunks without keys:

> *The notion that someone would market a closet that could never be opened—even if it involves a case involving a child kidnapper and a court order—to me does not make any sense.*[462]

278

A real-life excuse to break down those private doors came in December 2015 following a deadly terrorist attack that murdered fourteen people and injured twenty-two others at a City of San Bernardino, California holiday party. The assassins, Syed Farook and Tashfeen Malik were killed by police in a shootout two miles from the scene a few hours later.

The FBI soon collected whatever background they could to determine any particulars that might help them identify other potentially dangerous individuals who might be connected to the couple. They learned that twenty-seven-year-old Farook was the son of Pakistani parents who had immigrated to Illinois before he was born, making him a natural-born U.S. citizen.

Malik had lived in Saudi Arabia before meeting Farook during his hajj—pilgrimage to Mecca—and later coming to America. It turned out that she had pledged her allegiance to ISS on Facebook just before the attack, but this was not noticed until after it had occurred.[463]

The couple had worked hard to cover their electronic tracks before the attack—smashing personal phones and hard drives, deleting emails, and using a disposable burner phone. However, they had overlooked one informative detail that would reignite the encryption debate for months. Farook had left behind his work-issued iPhone 5C.

The FBI was eager to access all information Farook's iPhone might yield, including Facebook communications with any associates and, most vitally, his GPS coordinates just before the attack. Yet a big problem remained. All that Facebook information was locked up his phone with an encrypted code, and now he was dead.

Apple's iPhone safety features were designed to prevent anyone—including the FBI—from hacking into their equipment. Included is a software provision that wipes away all data after ten wrong password attempts. This feature is incorporated as a safeguard to protect users against any hacker who breaks into a phone—particularly criminals seeking financial information, credit card numbers, or information about how to gain access to a house or safe.

The San Bernardo terrorist attack investigation dilemma fit James Comey's case against personal data encryption perfectly. If there were other ISIS-inspired Americans or immigrants who were in communication with Farook and Malik, they needed to be picked up quickly. He insisted that if Apple would unlock the iPhone code, he would use that access with discretion.

Apple, on the other hand, argued that in the name of privacy and security, even they didn't know Farook's passcode. Nor would they agree to help the FBI.

In reality, David Singer reports that, according to a subsequent report by the FBI inspector general, the FBI may have already possessed the technology to unlock the phone. Senior FBI officials were desirous of confronting Apple in court on the matter.

Apple dug in on its resistance. CEO Tim Cook wrote a 1,100-word letter to his customers that was striking for its critical stance against the Obama FBI position which would sacrifice the privacy of American citizens.

Cook's letter stated, in part:

> *The United States government has demanded that Apple take an unprecedented step which threatens the security of our customers. We oppose this order, which has implications far beyond the legal case at hand.*
>
> *Some would argue that building a back door for just one iPhone is a simple, clear-cut solution. But it ignores both the basics of digital security and the significance of what the government is demanding in this case...The implications of the government's demands are chilling...ultimately, we fear that this demand would undermine the very freedoms and liberty our government is meant to protect.*[464]

Although Cook couldn't reveal it publicly, the FBI's demand that Apple breaks into the San Bernardino phone was just one of four thousand law enforcement requests to his company in the second half of 2015 alone.

Nevertheless, Comey wasn't about to back down: he told aides that the publicity around the San Bernardino case would, if anything, remind criminals, child pornographers, and terrorists to use encryption.

This was the moment, he said, to settle the encryption wars once and for all.

In the end, the FBI got into terrorist Syed Farook's encrypted iPhone Facebook passcode account without Apple's help or permission. According to Sanger:

> *The FBI paid at least $1.3 million to a firm it would not name—believed to be an Israeli company—to hack the phone.*
>
> *The FBI refused to say what the technical solution was, or to share it with Apple, apparently for fear that the company would seal up whatever hardware or software loophole was discovered by the hired hackers.*[465]

Facebook, by comparison, was more helpful in assisting the FBI investigators and French authorities in cracking a different terrorist group.

The first of three ISIS attacks began at 9:20 p.m. on November 12th 2015, when a suicide bomb went off outside the Stade de France in Saint-Denis. Shootings followed on the streets of Paris nine minutes later. Panicked outdoor café diners raced to the backs of restaurants for shelter.

Twenty minutes after the first attack, the shooters entered the nearby Bataclan Music Hall.

A few people in the back rows heard the assailants shout *Allahu*

Akbar as the song *Kiss the Devil* poured through theater speakers. That was just before the brutal assassins opened fire on the audience from the mezzanine level; then calmly—cold-bloodedly—walked up and down the aisles, randomly firing at anyone still moving.

By the time the siege of Paris ended, a little after midnight, 130 people had died. Two-thirds of the victims had been helplessly trapped in the theater's killing zone.

Paris police turned to Facebook to seek information from people around the world about identities and contacts of the nine ISIS members killed in the attack aftermath. It quickly became clear that several of the terrorists had multiple Facebook accounts that reflected their split lives. Some showed normal European lifestyles, while others, under norms de Guerre, portrayed lives of struggle against the West.[466]

French and the FBI authorities rapidly obtained court orders from previously notified allowing Facebook to legally turn over terrorist background data including a treasure trove of links between the accounts and specific cell phone numbers. David Sanger observes that the connections drawn so quickly from the Facebook community of ISIS supporters helped to dismantle the cell's support structure.

He writes: "No one will ever know how many lives that swift action saved."[467]

Finally, as for the general public, how have most Americans viewed the issue of government surveillance and privacy since first revealed in the Snowden leaks?

A 2018 Pew Research Center survey conducted five years after the releases revealed mixed results which are briefly summarized in the following:[468]

Americans were about equally divided regarding the impact of the leaks immediately following Snowden's disclosures, but a majority believed that the government should prosecute the leaker. Adults younger than 30 were more likely to believe that the leaks served the public interest.

Americans became somewhat more disapproving of the government surveillance program in ensuing months, even after then-President Obama outlined changes to NSA data collection. A majority (56%) didn't think that courts were providing adequate limits on phone and Internet data collected beyond anti-terror efforts.

By late 2014 and early 2015, disclosures about government surveillance had prompted some Americans to change the way they use technology. Among the 87% of U.S. adults who had heard of the programs, 25% reported to have "greatly" or "somewhat" changed at least one or more items including email accounts, search engines, social media sites, cell phones, mobile apps, text messages, and landline phones.

By 2014-2015, Americans broadly found it acceptable for the government to monitor certain people, but not U.S. citizens. About 82% of all adults said it was acceptable to monitor communications of American leaders and foreign leaders (60% each). Yet 57% believed it was unacceptable for the government to monitor communications of U.S. citizens.

About half of Americans in 2014-2015 expressed worry about surveillance programs of their own data. Roughly four-in-ten said that they were somewhat or very concerned about government monitoring their activity on search engines, email messages, and cellphones; about three-in-ten expressed similar concern over the monitoring of their activity on social media and mobile apps.

Few (only 9%) of those polled in 2014 said they had a "lot" of control over the information collected about them in their daily life...particularly with regard to how personal information is collected and used by companies.

Some 49% said in 2016 that they were not confident in the federal government's ability to protect their data. Only 12% were "very confident."

Americans had more confidence in other institutions, such as cell phone manufacturers and credit card companies. Around two-thirds were either very or somewhat confident about email providers keeping their information safe and secure.

In 2016, roughly half of Americans (49%) said their personal data were less secure compared with five years prior. Those ages 50 and older were somewhat more likely to express concern (58%) than those younger (41%).

Competing Infotech Agendas and Alliances

David Sanger, who teaches national security policy at Harvard's Kennedy School of Government, believes that China has closely watched legal struggles over privacy between government and social media platforms such as Apple, Facebook, and Google because "anyone with any technical skills knows that if you create an opening for the FBI, you create one for China and Russia and everyone else."

An Apple agreement to create a backdoor for the FBI, for example, would give the particular company no choice but to create one for China's Ministry of State Security too, or otherwise be ejected from the Chinese market.[469]

I elaborate many reasons in my book *"The Weaponization of AI and the Internet"* why collaborations between astoundingly powerful Silicon Valley information technology behemoths and Chinese overlords of population surveillance and control should greatly concern U.S. all.[470]

In an Oct. 1st 2018 speech, focused on security and economic issues at the Hudson Institute, Vice President Pence called on all U.S. companies to reconsider business practices in China that involve turning over intellectual property or "abetting Beijing's oppression."[471]

Pence said, "For example, Google should immediately end development of the Dragonfly app that will strengthen Communist Party censorship and compromise the privacy of Chinese customers."

Responding to Pence's criticism, a Google spokeswoman simply described the company's work as "exploratory" and "not close to launching a search product in China."

A logical follow-up question would be, "exploratory to what purpose?"

Meanwhile, we might also wonder why Facebook has explored what the New York Times referred to as "creepy patents" that will track "all most every aspect of its users' lives."

One patent application uses information about how many times you visit another user's page, the number of people in your profile picture and the percentage of your friends of a different gender to predict whether you're romantically involved with anyone.

Another Facebook patent application characterizes your personality traits and judges your degree of extroversion, openness or emotional stability to select which news stories or adds to display.

Facebook filed a patent application that reviews your posts, messages, and credit card transactions and locations to predict when a major life event such as a birth, death or graduation is likely to occur.

Facebook will be able to identify the unique "signature" of faulty pixels or lens scratches appearing on images taken on your digital camera to figure out that you know someone who uploads pictures taken on your device, even if you weren't previously connected.

They will also be able to guess the "affinity" between you and a friend based upon how frequently you use the same camera.

There is a Facebook patent application that uses your phone microphone to identify the electrical interference pattern created by your TV power cable to reveal which television shows you watched and whether you muted advertisements.

A couple of other patent applications will track your daily and weekly routines to monitor and communicate deviations from regular activity patterns, your phone's location in the middle of the night to establish where you live when your phone is stationary to determine how many hours you sleep, and correlations of distance between your phone's location and your friend's phones to determine with whom you socialize with most often.

As discussed in my book *The Weaponization of AI and the Internet,* monopolistic infotech behemoths with astonishing economic lobbying swag have come to exert tremendous influence over our open access to *all* public and private information and opinion discussions. Their status as corporate—rather government—entities, entitles them to determine what we are allowed to see, what we are not allowed to see, and from whom, entirely at their discretion. Moreover, they can do so without having to justify their specific-case decisions or to explain their rationale.[472]

Moreover, their wealth and global influence are growing at an astounding rate. According to PricewaterhouseCoopers, four of the world's ten largest market cap corporations are Silicon Valley companies: Amazon ($860 billion), Microsoft ($833 billion), Alphabet-Google ($765 billion), and Facebook ($445 billion). Two of the remaining ten largest are Chinese information tech companies; Tencent ($496 billion), and Alibaba ($370 billion).

British author and journalist, Steven Poole asks U.S. to consider how worried we should be regarding which vision prevails and who wields what influence over the broad personal and public aspects of our lives. Writing in *The Guardian* he urges that in contemplating our answer, we attempt to imagine the sort of lives we would be willing to accept for ourselves—or our children and grandchildren—a decade or two from now, given the trends we are already witnessing today.

Poole writes:

> *Imagine, for example, that as soon as you step outside your door—maybe even before—your actions are swept into a digital dragnet. Video surveillance cameras placed everywhere record footage of your face which becomes entered for correlation by feature recognition algorithms matched with photos on your driver's license to a*

national ID database.[473]

The reason offered for these intrusive intrusions on our privacy, of course, is that all of this is being done in your best interest to enhance your safety.

This being the case, Poole continues to question what this "best safety interest" answer has to do with allowing stealthy algorithms to keep track of what you purchase online; where you are at any given time; who your friends are and how you interact with them; how many hours you spend watching television or playing video games; and what bills you pay (or not).

These intrusions should be quite to imagine because they are already occurring.

As Rachel Botsman reminds us in her book *Who Can You Trust? How Technology Brought U.S. Together and Why It Might Drive U.S. Apart,* most of this already happens thanks to social media data-collecting behemoths. Ms. Botsman informs U.S. that a 2015 Office of Economic Development (OECD) study revealed that at that time there were already about one-quarter as many connected private and government-operated monitoring devices in the United States as the entire population.

Other agencies are likely to follow the NSA watchdog lead. Although the U.S. Department of Transportation Security Administration scrapped a proposal to expand PreCheck background checks to include social media records, location data and purchase history following heavy criticism, a major new terror incident can readily revive the plan.[474]

After all, what could possibly go wrong?

Maybe consider whether you would wish to endorse the adoption of the State Council of China's 2014 document called *Planning Outline for the Construction of a Social Credit System* as a credo for your nation's future. As the document claims:

> *It will forge a public opinion environment where*

keeping trust is glorious. It will strengthen sincerity in government affairs, commercial sincerity, social sincerity and the construction of judicial credibility.

In other words, trust the government to ensure everyone's sincerity on all matters, particularly sincere universal approval of the credibility of its omnipotent judicial sovereignty.

Even in the US, credit scoring is not certainly not a new idea not. More than 70 years ago, two Americans, Bill Fair and Earl Isaac, created the Fair Isaac Corporation (FICO), a data analytics company, to establish credit scores that can be applied by commercial companies to determine many financial decisions. These client service rankings include determinations regarding whether an applicant should be given a loan, and what interest rate should be applied to a consumer property mortgage.

Unlike the US, at least so far, there is little if any daylight in China between government and its high tech corporate agendas.

Paul Triolo, a technology analyst at the Eurasia Group think tank, warns of ominous global influences as Beijing leaders nationalize some of its high tech giants into de facto instruments of the state. China's so-called "Great Firewall" is effectively creating a dilemma of two Internets with separate communication privacy standards. The firewall lets the government control the web content Chinese citizens can view.[475]

Google's China-specific search engine, Google.cn, was launched in 2006 as a means for the company to stay in the country while abiding by its strict censorship rules. The company's normal search engine was also still technically available but was heavily filtered.

Critics charge for example, that China could use the new Dragonfly to replace air pollution information in online news reports, giving the appearance that pollution levels aren't as dangerous as they are.

Google eventually bailed on the project in 2010 amid China's accusations that the company allowed pornography on its search engine. A sophisticated hacking attempt on Google that originated in China was the last straw, at least temporarily prompting the company to stop censoring content thee and move its operations from the mainland.

While Google says it's only exploring a new search engine for the Chinese market, the tech giant already has an existing presence in the country. China is the world's most populous country with a growing middle class which is becoming increasingly connected. Abandoning China would cede the country's huge market to homegrown competitors like Baidu, China's largest search engine.

Nevertheless, U.S. lawmakers are understandably lambasting Google for even considering working with the Chinese government again. Senator Tom Cotton (R-AR) issued a statement in August 2018, calling out the tech giant for potentially working on a new Chinese information search engine. He wrote:

> *Google claims to value freedom and one hopes Google will put its corporate principles and America first, ahead of Chinese cash.*[476]

Chinese and other Internet-connected companies are being forced to make their businesses work uniformly everywhere else they can. An influential model for accomplishing this is the General Data Protection Regulation (GDPR) enacted in 2018 under European Union law to protect the privacy of all individuals within the EU and European Economic Area (EEA).

GDPR requires that all business processes that handle personal data must be designed and built with consideration of the principles and provide safeguards to protect data, using the highest-possible privacy settings to prevent the outside transfer to third parties without explicit, informed consent. The data also cannot be used to identify a subject without additional information stored separately.

Some U.S. tech companies are already doing this. Microsoft Corp. Said it will apply GDPR rules across all its services throughout the world. Apple Inc. has for years positioned itself as the data protection and privacy company, and its CEO Tim Cook supports a U.S. privacy law in line with Europe's. Facebook, on the other hand, has tried to make an end-run around EU rules by giving its users a stark choice: give up some rights or delete their accounts.

As reported by Kevin Williamson in *The Wall Street Journal*, Facebook has made it known that it would find it easier to operate under a single, international regulatory regime. He quotes CEO Mark Zuckerberg calling for "a more [globally] standardized approach" which is aimed vaguely at "protecting society" as a matter of thwarting "threats" to public safety." [477]

Williamson, also a correspondent for the *National Review*, notes that Facebook is already working with European governments to craft a regulatory regime it can live with. He comments:

> *That's troubling. Freedom House reports there is 'no official [Facebook] censorship'. In Austria, even as it admits that some speech, notably pro-Nazi political speech, is prohibited by law, which is the definition of official censorship. In Austria, possession of banned books can be punished with prison sentences of up to 20 years.*

Other more data-dependent U.S. companies aren't eager to comply with GDPR standards either. Google is fighting efforts to export those rules. The company's official wish list on "responsible" data-protection regulation includes a "flexible" definition of personal data and no restrictions on the geographic allocation of data storage. [478]

Google collaborated with Ascension, America's second-largest health system, to collect health information on 50 million U.S. patients scattered across 40 data centers in more than a dozen states without permission from either those insured individuals or their doctors. Code-named "Project Nightingale," the harvested data included personally identifiable names and birth dates, lab tests, medication and hospitalization history, and some billing claims and other clinical records.[479]

The 1996 federal Health Insurance Portability and Accountability Act (HIPAA) which generally allows hospitals to share data with business partners without telling as long as the information is used "only to help the covered entity carry out its health care functions." Personal privacy invasion issues and potentials for such information to be weaponized by government and corporate health care providers surrounding these practices raise a legitimate concern.

Some of the biggest tech companies currently have as much power over the hearts and minds of people as do the governments in countries where they operate. Moreover, individually and collectively, they also reach and influence more people than any other company have in history.

Citizens, bureaucrats, and politicians all over the world are now beginning to push back against that power.

Writing in *The Wall Street Journal*, Christopher Mims observes that as people become more familiar with the Internet, their views tend to change from enthusiasm to caution. This push-back against the power of digital giants also started in the West, where countries have been feeling the results of Big Tech's growing power the longest.

Mims cites a survey by the Centre for International Governance Innovation which reveals that in Kenya, for example, people are singularly positive about the impact of tech, whereas in North America and Europe, people are more concerned about overreach.

"Familiarity breeds contempt," said Fen Hampson, director of global security and politics at GIGI, who conducted the survey.[480]

The backlash is often directed at America's tech giants, such as Alphabet Inc.'s Google, Facebook Inc. and Amazon.com Inc., and how their ubiquity affects individuals and businesses. This sort of push-back will benefit burgeoning tech industries in some big countries such as China and India which can continue to dominate their large domestic markets. Tech services in many smaller countries that can't compete with the giants will be left to negotiate business roles and conditions wherever they can.

Christopher Mims views competition between large and small domestic tech companies as one that can subdivide Internet in a way that will force the biggest players to create separate products and procedures for different regions. He foresees that the results—following a costly, complicated and protracted transition—will be better for consumers in some cases, and significantly worse in others.[481]

Facebook, Twitter and their social media peers were at one time positively credited with spearheading pro-democracy uprisings that toppled dictators throughout the Middle East. Their services were also seen as a great benefit during natural catastrophes, allowing authorities the means to convey crucial information, and to organize assistance.

Those same social media sites are increasingly being used and accurately perceived as forces that corrode democracy as well as to promote it. Tyrants and terrorists exploit their vast public outreach capacities to spread disinformation and to fuel ethnic violence.

At the same time, and as Emma Llansó, the director of the Center for Democracy and Technology's Free Expression Project notes, censorship laws can serve as a "pretext for enforcing against political dissidents or journalists." [482]

Facebook and Google remain banned in China. YouTube has been periodically blocked in more than two dozen countries since the service was founded, including an incident in 2007 when a Turkish court ordered the removal of videos critical of the country's founder. Russia has recently sought to criminalize the spread of "fake news."

Such censorship is said to be justified as *streitbare Demokratie—militant democracy,* a German constitutional principle under which certain kinds of political communication are censored, and certain political parties and organizations are prohibited. This theory holds that liberal democracies must sometimes act in illiberal and antidemocratic ways to counter political tendencies that constitute threats to the liberal democratic order.[483]

Writing in the *Wall Street Journal*, Kevin Williamson warns U.S. to recognize erosions of our American First Amendment free speech principles as this theory becomes weaponized and vulgarized under the slogan *"No free speech for Nazis."* The term "Nazi" can be ubiquitously applied to encompass almost everyone—ranging from radical feminists with unfashionable ideas about transgender issues to mild-mannered fellows of the American Enterprise Institute.

Williamson argues that although European practices are significantly at odds with America's First Amendment, they are made somewhat more understandable when viewed in a postwar European context, one in which a revanchist Nazi movement wasn't previously unthinkable at all. He asks U.S. to consider that the dicey part in all of this is trying to figure out what constitutes a genuine threat:

> *Is it a group of skinheads in Munich planning a pogrom? A bookseller stocking 'Mein Kampf'? An anti-Semitic letter from Henry Ford?*[484]

As a practical matter, Williamson predicts that eventual and perhaps inevitable government standards governing speech restrictions on tech platforms will reflect more restrictive European practice rather than more liberal American practice. He points out that the Big Tech companies will ultimately jump in line for much the same reason that California's relative stringent automotive emission standards act as an effective national standard: "Corporations generally prefer standardization and homogenization where they are economical."

This being the case, Kevin Williamson urges U.S. to be clear about regulatory standardization and homogenization means: Official censorship by governments abroad, proxy censorship by business interests at home, and doublespeak on all sides. He prudently concludes:

> *You can have streitbare Demokratie, or you can have a political culture so childish and illiterate that it can't distinguish between Charlie Kirk, Charles Murray and Charles Mansion. But you can't have both and a free society too.*[485]

More encouragingly, In the spring of 2018, about three dozen companies—Microsoft, Facebook, and Intel among them—agreed to the most basic set of principles, including an innocent-sounding vow that the signatories would refuse to help any government, including the United States, mount cyberattacks against "innocent civilians and enterprises from anywhere."

The companies also committed coming to the aid of any nation on the receiving end of such attacks, whether the motive for the attack is "criminal or geopolitical."

David Sanger observes that while this was a start, it was a barely satisfying one. No Chinese, Russian, or Iranian companies were part of the initial compact, nor were some of the biggest forces of the technology world, including Google and Amazon, both struggling between their desires to do vast business with the U.S. military and their desires to avoid alienating their customers.[486]

Sanger compared the first principles put forth in the agreement as being like the Internet—"sprawling and messy". No mention was made of supporting human rights—meaning that if Apple it later joined the accord, could still get away with its decision to bow to Beijing by keeping its data on Chinese customers on servers inside China.

If there is a lesson that emerged from years of trying to find, follow, and disrupt terrorists, it is that the same infotech overlords that figured out how to destroy centrifuges and disrupt power grids from afar have also learned how to control and weaponize social media platforms.

For years, the world's most brilliant Silicon Valley technologists convinced themselves that once they connected the world, a truer, global democracy would emerge. Over time, however, a harsher possibility has emerged a realization that each of U.S. must take on more responsibility to protect ourselves.

Sanger points out that we need to remember that the same technologies we humans create to enrich our societies and lives can also be used to plunge both ourselves and our adversaries into darkness. The good news is that because we created these technologies, we also have chances of controlling them to manage the risks.

Many of these risk-management strategies might seem quite obvious and relatively simple. For example, we can be more aware of what phishing campaigns look like, how to lock up home-network Wi-Fi routers, and how to sign up for two-factor authentication.

Sanger estimates that these steps can help to wipe out about 80 percent of our daily threats. Just as we wouldn't leave our doors locked when we leave home or our keys in the ignition of our cars, we also shouldn't leave our lives exposed on our phones, either.

He concludes: that just as this has worked in other realms, it can work in cyberspace as well.[487]

CYBERSECURING AMERICA'S FUTURE

BROOKINGS INSTITUTION RESEARCHERS Peter Singer and Alan Friedman characterize the cyber world as a place of exponentials, a world where the continuous expansion of online information multiplies upon itself year after year at a growth rate equivalency of online threats.

There is one piece of this world, however that works at exponential speed: the government. Instead, they argue, "it moves at a glacial—if that." [488]

Rethinking Government-Private Roles

Singer and Friedman note that in 2004, the U.S. Government Accountability Office (GAO) identified a comprehensive set of characteristics that the U.S. executive branch needed in a national cybersecurity strategy. The proposal topics ranged from allocating federal resources to defining accountability policies.

A full decade later, the GAO reported that the White House

was essentially at the same point, stating,

> *No overarching cybersecurity strategy has been developed that articulates priority actions, assigns responsibilities for performing them, and sets timeframes for their completion.*

Congress, although purporting to be interested in cybersecurity, did little but endlessly talk about it. Despite holding as many as sixty hearings a year on the subject, not a single piece of cybersecurity legislation was passed between 2002 and more than a decade later.

Writing in 2014, Singer and Friedman conclude that government efforts had produced a patchwork of agencies and projects, often with a little clear strategy and mixed levels of control:

> *For instance, the NSA and DoD have worked together to share attack signatures with a group of critical defense contractors, while NSA agreed to offer technical support to Google following attacks in 2010 and to the financial industry following a series of DDoS attacks in 2012.*[489]

Singer and Friedman also points out that mixed government agency and their outsourced intelligence organizations have created contentious conflicts between national security versus personal privacy. A long trail of public scandals runs from the NSA's Prism and Verizon scandal in 2013 (as revealed by the Edward Snowden leaks); to the 2005 controversy over warrantless surveillance programs ordered by the George W Bush administration; to covert political CIA and NSA roles in the illegal Obama administration surveillance of President Donald J. Trump.

Big problems arise as various and diverse government regulatory agencies have overlapping and competing for security roles, and gaps in authority.

The U.S. Department of Homeland Security's Computer Emergency Response Team (US-CERT), for example, is often aided by the National Institute of Standards and Technology (NIST) located in the Department of Commerce. NIST is tasked to work with industry to develop and apply technology, measurements, and standards in everything from weights used at grocery stores to the primary building blocks of information systems.

Making matters even more confusing, While NIST also shares responsibility with the North American Electric Reliability Corporation(NERC) for developing Smart Grid standards, neither has an explicit responsibility to lead security initiatives. As a result, the distribution layer of the power grid is not covered by either entity, creating a situation where two agencies simultaneously both have—and also do not have—the ability to set security standards.[490]

This regulatory confusion has created frequent conditions whereby firms either aren't held to standards or have to sort out what regulations must be complied with. In addition, it dilutes the ability of any single agency to affect meaningful security upgrades, creating dangerous regulatory gaps that bad guys can hide in.[491]

Whereas there are broad public and private sector agreements regarding the critical importance of cybersecurity standards, there is considerably less consensus about where and how much government engagement and authority moth most appropriate and effective. Many corporate entities and their lobbies legitimately resist government regulatory overreach which adds substantial operation costs while providing little or no real security benefits.

Several major American power companies argue the known loss of revenue needed to take plants offline for just a few hours to upgrade their cyber systems is greater than any unknown cyber risks, which they are not sure they face or would even be defeating.

The U.S. Chamber of Commerce has expressed similar concerns that bureaucratic regulatory agencies shift "businesses' resources away from implementing robust and effective security measures and toward meeting government mandates." [492]

Peter Singer and Alan Friedman observe that such tensions between the private sector and public interests neatly flips for attacks against critical infrastructure:

> *Essential industries make the case that national defense is a public good, and therefore they should not have to bear the costs of defending against cyberattacks of a political nature, any more than they should have to provide their own anti aircraft guns to defend against an enemy's bomber planes.*
>
> *Meanwhile, the government has to worry that the public is dependent on infrastructure like power and water plants, where the owners see little incentive in paying real money to secure facilities against a risk that can't be stated at the bottom of a monthly business report.* [493]

Even though cybersecurity remains a major public concern, many, if not most decisions to secure it are controlled or influenced by private actors, including all of U.S. who decide whether or not to click on a link. Others are companies and individuals who determine whether or not to invest in security, and how; along with technology vendors and the creators whom they buy from that expose and protect clients and their operating systems from vulnerabilities. [494]

The number and variety of these technology equipment and service suppliers are large. Adding to the security challenges, devices on the Internet ranging from computer terminals and laptops, routers and switches, email and web page servers, and data files are often produced separately from companies that run these devices.

In the US, market, most laptops are made by Dell, HP, and Apple. (A Chinese company, Lenovo, is making a dent after having bought IBM's laptop computer unit.)

Servers are made by HP, Dell, IBM, and a large number of

others, depending on their purpose. The software they run is written mainly by Microsoft, Oracle, IBM, and Apple, but also by many other companies.

Most big routers are made by Cisco and Juniper, and now the Chinese company Huawei.[495]

The numerous elements and interfaces of connected networks present a wide spectrum of cybersecurity challenges. Singer and Friedman characterize different types of threats according to *low, medium,* and *high* classification categories:

> *Low threat level potential is malware able to affect systems from the outside, but unable to penetrate the target or to create direct harm. In this category, there are tools and software designed to generate traffic to overload a system and adversely impact its services with a temporary effect (for example, Denial of Service attack) without actual software or hardware damage.*
>
> *Medium threat level potential is any malicious intrusion that is able to disrupt or modify the behavior of systems and steal information but cannot result in kinetic harm to a person. Generic intrusion agents such as malware, which can spread rapidly, are included in this category.*
>
> *High threat level potential is an agent that is capable of penetrating the target, avoiding security controls (antivirus) and creating direct kinetic harm to the victim (equipment or person). An example is the Stuxnet virus which targeted and harmed a specific cyber-physical system. Such sophisticated viral types include those with and without self-learning capabilities.*[496]

Recognizing that ever more sophisticated cyber methods and

malware will sometimes succeed, the second level of defense involves resilient damage control to preserve vital organization operations, data, and network system functions.

Resiliency fundamentally requires expecting the unexpected, understanding how all vulnerable targets fit together, and planning how to maintain continuity of vital functions when under attack.

Typical preparedness contingency actions ensure abilities to rapidly detect and shut down a violated device or system, lock-down valuable information files and transfer basic system operations to more secure back-up mechanisms.

Analyst at Lockheed Martin outlined a series of stages by which hackers bring off a successful attack. This "kill chain" moves from "reconnaissance and weaponization" to "delivery", exploitation, installation, command and control, and actions."

To prevent losing data, a resilient defender need only detect an incursion at any one of these stages.

Although major corporations are accustomed to providing business and service continuity plans for natural disasters and accidents, cybersecurity—by its devious human nature—is far more complicated, geographically expansive, and unpredictable. Whereas traditional localized precautions focused primarily upon predictable risks, cyberattacks can come from anywhere without severe weather or element failure warnings.[497]

Former U.S. presidential counter-terrorism officials Richard Clarke and Robert Knake believe that while there is no way to block incursions from China, Russia or Iran, there are still well-proven strategies for limiting damage from these hacks. They report that many smart companies already use hundreds of different tests to remove hackers, including tests based on what cyber attackers do once they get inside a network.[498]

For example, healthcare giants UnitedHealth and Aetna have avoided big data losses, while one of their rivals, Anthem (the second-biggest healthcare company), lost the records of all its customers in 2015 when it failed to encrypt their personal data.

Nevertheless, these security efforts come at a cost. JPMorgan Chase, the country's largest bank, spent $600 million for information security in 2017. For smaller companies, information-sharing allows for "herd immunity"—where other companies in a group are alerted as soon as malware is discovered at any one of them.

David Sanger argues that while the top tier of the financial industry and electric utilities have made huge investments to safeguard their networks against hackers, smaller banks and rural power companies become more vulnerable targets.[499]

Corporations, large and small, are not the only rapidly growing cybersecurity concern.

As we continue to put autonomous cars on the road, connect Alexas to our lights and our thermostats, attach Internet-connected video cameras on our houses, and conduct our financial lives over our cell phones, each and all of our vulnerabilities expand exponentially.

And the very same Internet vulnerabilities apply to those government entities that we count on to protect U.S. from harm.

Stiffening Our Internet Backbone

An estimated 98 percent of U.S. government communications, including classified military communications, travel over civilian-owned-and-operated networks and systems. While there are hundreds of independent service providers, the half a dozen or so large Tier 1 ISPs constituting the Internet backbone include AT&T, Verizon, Level 3, Qwest and Sprint. These "trunks," in turn, connect to other smaller ISPs within the country.[500]

Given the predominantly private nature and control of the Internet, government involvement in its policies, even for vital security purposes—isn't universally welcomed.

John Barlow of the Electronic Frontier Foundation perhaps captured this sentiment best in his *Declaration of the Independence of Cyberspace.*

Governments of the Industrial World, you weary giants of flesh and steel...I ask you of the past to leave U.S. alone. You are not welcome among us...You have no sovereignty where we gather...Cyberspace does not lie within your borders...You claim there are problems among U.S. that you need to solve. You use this claim as an excuse to invade our precincts. Many of these problems don't exist. Where there are real conflicts, there are wrongs, we will identify them and address them by our means.[501]

Peter Singer and Alan Friedman point out two problems with this way of thinking:

First, governments have seen the Internet become crucial to global commerce and communication, but even more so to their own national security and economic prosperity. And so, even if they're unwanted, they are involved, simultaneously solving and creating real problems in cyberspace.

Second, while there may be no actual physical territory or borders in cyberspace, every node of the network, every router, every switch is within the sovereign borders of a nation-state, and therefore subject to its laws or travels on a submarine cable or satellite connection owned by a company that is incorporated in a sovereign nation-state and therefore subject to its laws. In other words, there is no non-sovereign, 'free' part of cyberspace.[502]

General Keith Alexander, head of both the NSA and U.S. Cyber Command, has argued for a "secure, protected zone" within the

Internet. Similarly, the FBI's former Assistant Director Shawn Henry argued for a "new, highly secure alternative Internet."

Dividing or breaking up the Internet backbone into smaller parts presents conflicting security versus utility tradeoffs. Generally speaking, network protection is generally inversely correlated with size, while network utility is positively correlated.

Put another way, the bigger the network, the more security problems; the smaller the network, the less useful it is. Networks that span large sets of organizations open up additional numbers and types of risks, including human errors and equipment and software vulnerabilities.

Singer and Friedman urge that we in the U.S. must set aside a separate more secure section of cyberspace with a new region of peace and predictability within the lawless Wild West of the Internet to protect highly critical infrastructure systems, such as power plants along with most frequent online targets, such as consumer banks.

Singer and Friedman propose that a new backup national power grid might be established which is entirely unconnected to the Internet. Without it, they say, the U.S. is defenseless against "somebody like the Russian GRU, engaging in a cyberattack that would technically revert U.S. to the nineteenth century, but without all the equipment that people in the nineteenth century had to deal with life in a society without electricity." [503]

America's national security agencies and the public have good reasons to worry about logic bombs since they seem to have found them all over our electric grid. Nevertheless, establishing a government-wide model as the solution would also likely need to include access for many smaller organizations such as protections for rural electricity providers.

Also, the idea of entrusting government with an expanding foothold of control over the Internet of Things would need to ensure rigorous and independent personal privacy oversight to prevent Big Brother from illegally spying on us. Moreover, the same concerns holds true regarding privacy invasions on behalf of the backbone Tier 1 and smaller ISPs.

In any case, significant security problems remain to be addressed regarding screening outside malware-infected traffic before it enters the national Internet backbone.

The independent Federal Communications Commission (FCC) currently has the authority to issue regulations requiring the Tier 1 ISPs to establish such a protective system. Nevertheless, ISPs who have repeatedly known they were infected by botnets, haven't informed their customers (much less, cut off access) out of fear that they would switch providers and perhaps sue them for privacy violations.[504]

ISPs should be legally required to inform users when data shows that their network computers have been compromised.

Singer and Friedman point out that inspecting all the Internet traffic about to enter the backbone theoretically presents two significant problems, one technical and one of policies:

> *The technical problem is, simply, this: there is a lot of traffic and no one wants you to slow it down to look for malware or attack scripts. The policy problem is that no one wants you reading their emails or web page requests.*
>
> *The technical issue can be overcome with existing technology. As speeds increase, there could be difficulty scanning without introducing delay if the scanning technology failed to keep pace.*[505]

Several companies, in fact, are already demonstrating combinations of hardware and software capable of scanning Internet information traffic so fast that they do so without introducing any measurable delays in fiber optic "line rate." This absence of delay referred to as "no latency," passes the technical hurdle—at least for now.

As for the "policy problem," preventing government or ISPs

from reading our emails, Singer and Friedman believe that this issue can be resolved as well. They propose that an automated system used for "deep-packet" malware inspection would be *not* looking for keywords, but rather would only check for predetermined signature patterns of ones and zeros that match up with known attack software. Once discovered, the system would "back hole" to "kill" the packets, quarantine them, and set some aside for special analysis.[506]

The deep-packet inspection apparatus proposed by Singer and Friedman would be placed at "peering points" where fiber optic cables connecting to the U.S. Tier 1 ISPs come out of the ocean and connect to each other as well as to smaller networks. They suggest that the U.S. government—Homeland Security perhaps, would be a paying customer and that the networks would be run and managed by the ISPs and/or systems integrator companies.

The Department of Homeland Security's "Einstein" system has been installed at some locations where government departments connect to the Tier 1 ISPs. Einstein only monitors government networks.

The Department of Defense has a similar system at the sixteen locations where the unclassified DoD intranet connects to the public Internet.[507]

Private Internet security companies such as Symantec and McAfee currently have elaborate global systems that look for malware signatures, just as government and ISP experts do. In this regard, all would continue to work together to share information using "out-of-band communications" (not on the Internet) to maintain their security.

Verizon and AT&T already deploy deep-packet inspection capabilities at some locations which, as previously mentioned, have turned up malware signatures that were not reported to customers. At the same time, they are also often reluctant to black hole malicious traffic due to concerns of customer lawsuits over service interruptions.

Pooling government and private sector network capabilities and services in combination with ongoing advancements in higher speed capacity, more memory, processing capabilities, and out-of-band connectivity would shore up the U.S. side of the Internet backbone upon which all other national networks rely.

We must remember, however, that the Internet is global—not just national—and that this fact connects U.S. cybersecurity to much broader and more complicated international and intergovernmental policy issues.

The Internet traverses many world regions and visions of Internet governance and sovereignty with predominant contentions between authoritarian and democratic nations.

Whereas authoritarian regimes—China, North Korea, and Iran for example—seek to restrict control over their national bits of the Internet, democrat states regard Internet openness as a key to its successes and greatest value. Singer and Friedman observed that it is this open feature allows the Internet to evolve to meet its users' virtual wants and needs, regardless of their physical location in the world.[508]

Fundamental differences between authoritarian and democratic visions and priorities make for huge challenges in establishing meaningful and enforceable worldwide cybersecurity agreements.

Various proposals to create a set of globally accepted understandings—patterned after the 1923 Hague Rules of Aerial Warfare 1929 and/or Geneva Convention or 1923—for example, may put American security at greater risk:

Advanced cyber powers, including the United States, fear that such formal agreements will tie their hands from developing new and more effective weaponry and defenses while others either catch up or simply ignore the new laws.

Back in 2009, Russia floated the idea at the United Nations that the use of any cyberweapon by a state be preemptively banned, essentially trying to apply the model of arms control to cyberspace. Meanwhile, Russia has routinely been using non-state patriotic

Larry Bell

hacker networks to conduct cyberattacks.

The U.S. tends to view Wild West's behavior on the Internet as akin to problems in the original American West; a lawless environment of theft and bad guys running amuck with no consequences. This dangerous circumstance is made evident in espionage by Chinese and other foreign actors that target intellectual property and critical civilian infrastructure.

China and Russia, by contrast, often view Wild West behavior as the Western nations who export their democratic values. Clear evidence of this concern is illustrated by successful pressures upon U.S. Internet search companies to remove references to the 1989 protests in Tiananmen Square as conditions to maintain business offices in China.509

Securing Critical Infrastructures

The U.S. and other of the world's most economically and technologically developed nations are in a real sense more vulnerable to Wild West Internet and other cyber lawlessness than those countries with less to lose.

Former White House counter-terrorism officials Richard Clarke and Robert Knake ask U.S. to put aside, for a moment, the question of how a cyberwar might start, and consider a US-Chinese conflict as an example.

Although much of China is highly advanced, a lot of the country remains far less dependent than does the U.S. upon networks controlled in cyberspace. The authoritarian Chinese government will also have to worry less about temporary inconveniences experienced by its citizens, or become concerned about the political acceptability of measures it might impose in an emergency.510

Clarke and Knake urge U.S. to consider that while might have better offensive cyber weapons than others, the fact that we might be able to turn off the Chinese air defense system will give most Americans limited comfort if in some future crisis the cyber warriors of the People's Liberation Army have kept power off in

308

most American cities for weeks, shut the financial markets by corrupting their data, and created food and parts shortages nationwide by scrambling the routing systems at major U.S. railroads.

The increasing vulnerability of vital U.S. systems will continue to tempt opponents to attack in a period of tension when they think that they have an opportunity to reshape the political, economic, and military balance by demonstrating to the world what they can do to America. Their weapons to cripple the U.S. may already be in the US. And as Clarke and Knake note, "They may not even have entered America through cyberspace, where they might be discovered, but rather on CDs in diplomatic pouches, or in USB thumb drives in business men's briefcases." [511]

The most devastating cyber-attack nation-state can launch today to achieve a major impact on the U.S. would be to shut down sections of the Eastern or Western Interconnects, the two big grids that cover the U.S. and Canada. (Texas has its own third, grid).

Many people dismiss the significance of an attack on a power grid...blackouts from lightning storms, for example, are relatively common. But failure as a result of intentional activity will likely be much longer.

In what is known as the "repeated smack-down scenario," cyberattacks take down the power grid, and keep it down for months. If attacks destroy generators, replacing them can take up to six months because each must be custom-built.

Having an attack take place in many locations simultaneously, and then happen again when the grid comes back up, could cripple the economy by halting the distribution of food and other consumer goods, shutting down factories, and forcing the closure of financial markets.

How vulnerable is the U.S. power grid to cyberattacks?

Audits of six major cybersecurity firms to determine how long it took them to test abilities to hack into grid controls of power companies from the internet, all reported that no penetration had taken longer than an hour. That hour was spent hacking into the company's public website, then from there into the company's intranet, then through "the bridge", they all have to their control systems.[512]

Some of those audits cut the time by hacking into the Internet-based phones (voice over Internet protocol, or VOIP, phones) that were sitting in control rooms. These phones are by definition connected to the Internet; that's how they connect to the telephone network. If they are in the control room, they are also probably connected to the network that runs the power system.

In some places, the commands to electrical grid components are sent in the clear (that is, unencrypted) via radio, including microwave. Just sit nearby, transmit on the same frequency with more energy in your signal than the power company is using, and you are giving the commands (if you know what the command software looks like).[513]

Richard Clarke, an admitted President Obama advocate, believes that the Obama administration's "Smart Grid" plan (to make the grid even more integrated and digitized) would make it less secure. Instead, Clarke proposes the first step to make the U.S. national electrical system both smart and secure would be to issue and enforce serious regulations to require electric companies to make it next to impossible to obtain unauthorized access to the control network for the power grid.

Doing this would mean no pathway at all from the Internet to the control systems of hundreds of generation and transmission companies throughout three North American power-sharing entities.

Accomplishing this, however, would require federal regulation—a hard sell proposition to the power companies. When asked what assets of theirs were critical and should be covered by cyber regulations, the industry replied that 95 percent of their assets should be left unregulated with regard to cybersecurity.[514]

Clarke proposes that new federal cybersecurity regulations

should focus on disconnecting the control network for all power generation and distribution companies from the Internet, and; then making access to those networks requires authentication.

Authenticating the commands would mean that through a proof of identity procedure, or electronic "handshake," the generator or transformer would know for sure that the command signal it was getting was coming from the right place.

Then, just to make things even harder for an attacking cyber warrior, regulations should require that the actual control signals sent to generators, transformers, and other key components be both encrypted and authenticated. Encrypting the signals would mean that even if someone could hack their way in and try to give alternate instructions to a generator, they wouldn't have the secret code to do so.

As for the costs for implementing his proposal, Clarke says: "It would really not be all that expensive, but try to tell that to the power companies." [515]

Clarke and Knake conclude that while it is hopeless to try to protect every computer in the U.S. from cyberattacks, it may be possible to sufficiently harden the most important networks a nation-state would target:

> *We need to harden them enough then no attack could disable our military's ability to respond or severely undermine our economy.*
>
> *Even if our defense is not perfect, these hardened networks may be able to survive sufficiently, or bounce back quickly enough, so that the damage done by an attack would not be crippling.* [516]

The War Against Cybercrime

Peter Singer and Alan Friedman observed that government and

private security professionals who are tasked with defending advanced networks reportedly spend vastly more time, effort, and money addressing generic problems like botnets, spam, and low-level worms that hit all users on the Internet than they do on the APTs that hold the potential for far greater harm.[517]

Lines of when and how a cyberattack becomes an act of warfare or criminal activity and who can and should be held accountable for it remain fuzzy. Previously discussed "patriotic hackers," a gray zone exploited by some nation-states, such as in the Russian attacks on Estonia, offer a textbook example.

An ironic twist of patriotic hacking executed by citizens or groups upon enemies of their state is that their government which often plays an orchestrating role can later act as an aggrieved victim whenever accused of involvement.

Such prosperous arrangements afford criminal groups the freedom to operate in exchange for demonstrating their "patriotism" when governments ask for aid. Singer and Friedman suggest that we think of this as the cyber equivalent of the deal struck between the FBI and Mafia during World War II, when the Feds agreed to lay off their investigations in exchange for the mobsters watching the docks for Nazi spies and aiding military intelligence operations in Italy.

Patriotic hackers also provide government sponsors with supplementary expertise and resources that lie beyond the state.

In 2005, the Chinese military reportedly organized a series of regional hacker competitions to identify talented civilians. As a result, the founding member of the influential hacker group Javaphile (hugely active in Chinese patriotic hacker attacks on the U.S. in the early 2000s, including the defacement of the White House website) joined the Shanghai Public Security Bureau as a consultant.

Government, civilian state-sponsored and independent criminal hackers hallmark common cyberattack tools and methods and defense countermeasures. As such, whether enacted to gain a political objective, perpetrated for intellectual property espionage, or conducted as theft of monetary nature, the term "cybercrime" blurs singular motivational distinctions.

In all instances, cybercrime is most often defined as the use of digital tools by criminals to steal or otherwise carry out illegal activities. As information technology grows more pervasive, however, it becomes harder to find crimes that *don't* have a digital component.

The most pervasive type of cybercrime is "credential fraud", or the misuse of account details to defraud financial and payment systems. Such systems include credit cards, ATM accounts, and online banking accounts.[518]

To access these accounts, criminals can obtain credentials like passwords and other data wholesale by attacking computers that store account details for the merchants, banks, and processors in charge of the whole system.

Alternatively, the cyber thieves often go directly after the individual account owner by tricking him or her through a phishing email that poses as a communication from a financial institution and presents a link where the victim is prompted to enter his credentials.

In the end, a credential's worth depends entirely upon the criminal's ability to extract value. A credit card number, for instance, is only useful if the attacker can turn it into desired goods and services without being traced back to them.

Larger online banking frauds draw upon large organizational criminal infrastructures of sellers, resellers, patsies, and "money mules" who act as intermediate steps in the transfer of money or goods. This arrangement is similar to more traditional money laundering but in reverse. Rather than washing criminal profits into a wider legitimate pool, the goal is to create a series of seemingly legitimate transactions to get money into the dirty hands of the criminals.[519]

Cryptocurrency exchanges provide rich new cyber targets.

In November 2019, nearly $50 million was swiped from a South Korean cryptocurrency exchange called Upbit where 342,000 "ether" coins were sent to an unidentified crypto wallet. Ether, which runs on the Ethereum blockchain, is the second-largest cryptocurrency in market value, trailing only bitcoin.

Bitcoin, ether and other cryptocurrencies exist on independent networks and operate on blockchain technology, an open-ledger system that stores a public record of transactions. In an effort to replicate the anonymity of physical cash, those transactions aren't connected to an identity. The anonymity feature is appealing to proponents, but is also attractive to hackers and makes it difficult to catch thieves.

According to CipherTrace, a cryptocurrency research company, about two-thirds of the world's top 120 cryptocurrency exchanges reportedly have weak policies for monitoring customers on their respective platforms. This makes it easier for hackers—and other bad actors, such as money launderers—to misuse exchanges. An estimated $4.4 billion was stolen in cryptocurrency heists in 2019 through the month of September.[520]

Many cybercrimes target specific businesses more directly. Singer and Friedman report that by 2013, an average firm of 1,000 employees or more was spending roughly $9 million a year on cybersecurity, whether it was a bank or paint maker.[521]

Whereas cybercriminals first relied primarily on accessing widely available hacking tools, many are now starting to write their specially-designed code to beat security systems. This occurred, for example, in the theft of millions of credit card numbers from T. J. Maxx in 2003.[522]

New criminal enterprises have emerged as cybercrime laboratories where malicious malware payloads and exploits used in cyberwarfare are developed and refined. These trends evidence rapidly growing sophistication of cybercriminals that could potentially grow to become as sophisticated as the state-level threat.

The additional blurring of threats posed by cybercrime, cyberwar, and even aerial and field warfare will come to fruition as more and more digital software systems and equipment devices associated with both become fully embedded in the Internet of

314

Things.

Singer and Friedman recount an illustrative example that occurred in 2009 after American soldiers had captured an insurgent leader in Iraq. As they went through the files on his laptop computer, they made a remarkable discovery: he'd been watching them watch him.[523]

A key part of that U.S. military effort was to track movements of the insurgent ground force using a fleet of unmanned drone aircraft that beamed back video. But inside the captured leader's laptop were "days and days and hours of proof" that digital feeds were being intercepted and shared among various targeted groups.

The insurgents had evidently figured out how to hack and watch the drones' feed, like robbers listening in on a police radio scanner. Even more disturbing was that they used commercially-available software originally designed by college kids to illegally download satellite-streamed movies.

"Skygrabber," as the software program was known, cost only $25.95 on a Russian website.

Nevertheless, as Singer and Friedman point out:

> *Being powerful means you have a choice. Being weak means you don't. The insurgents in Iraq would rather have had the drones than their pirated video feed. That's why it still pays to be the stronger in battle, even on a cyber battlefield.*[524]

Asserting Proactive Global Leadership

The cyber era has dramatically changed challenges and necessities for America to project dominant "strength" against a growing number of technically-emboldened adversaries.

Possession of devastatingly potent weaponry is no longer restricted to the wealthiest nations. Russia, for example, despite— and likely as a result of its relatively anemic economy—is ramping up cyberwarfare capabilities that have been beta-tested in Ukraine and numerous other countries.

Economically-impoverished North Korea actually enjoys a cyber defensive advantage in that unlike America, the technologically backward country lacks dependence upon a highly vulnerable power grid and advanced information infrastructure.

China is also less dependent on a wired-together Internet of Things energy and information utilities than in the U.S. and is both stealing and investing in advanced cyberwarfare-applicable AI and computational technologies at a colossal pace.

Unlike past arms races, it is impossible to confidently assess an adversary's capabilities in order to plan and prepare comparably assured defenses. Similarly, unlike in conventional war, a superior offensive capability cannot be certain to detect and destroy all of an opponent's offensive capabilities.

Cyber adversaries often prioritize "soft targets," typically those with great potential to disrupt social and economic order short of triggering major military reprisals. As Clarke and Knake warn, the next U.S. war will be provoked by a cyberattack—perhaps one that misjudged escalation potentials and grossly underestimated the reprisal consequences.

David Sanger, who teaches national security policy at Harvard's Kennedy School of Government, recognizes that the U.S. government shouldn't respond to every cyberattack. In doing so America would be an inconstant low-level war. Also, not every cyberattack warrants a cyber response. Criminal attacks, Sanger believes, should be handled as other crimes are handled—with vigorous prosecution.

When government does respond to a major state-sponsored hacking event, Sanger advocates that we do so very publicly in a manner that will deter future attacks—as public as an American airstrike on a chemical weapons plant in Syria, or an Israeli airstrike on a nuclear reactor:

Every time we respond quietly—or not at all—to an attack because we are worried about revealing the quality of our detection systems or the capability of our weapons, we only encourage escalation and further cyber strikes from our adversaries.[525]

In addition to showing other nations that there will be a painful price to pay for truly serious cyberattacks, Sanger also emphasizes the importance of publicly clarifying to other nations—at the presidential level—that some things are off-limits. This policy statement must include what America commits not to do. Otherwise, he argues, there is no real hope of getting other countries to limit themselves accordingly.

Sanger urges that America's cybersecurity decisions be guided by five key rules:

- **Rule One:** We must recognize that our cyber capabilities are no longer unique. Russia and China have nearly matched America's cyber skills; Iran and North Korea will likely do so soon, if they haven't already. We have to adjust to that reality. Those countries will no sooner abandon their cyber arsenals than they will abandon their nuclear arsenals or ambitions.

- **Rule Two:** We need a playbook for responding to attacks, and we need to demonstrate a willingness to use it. It is one thing to convene a 'Cyber Action group,' as Obama did fairly often, and have them debate when there is enough evidence and enough concern to recommend to the president a 'proportional response.' It is another thing to respond quickly and effectively when such an attack occurs.

- **Rule Three:** We must develop our abilities to attribute attacks and make calling out any adversary the standard

317

response to cyber aggression. The Trump administration, in its first eighteen months, began doing just this; it named North Korea as the culprit in WannaCry and Russia as the creator of NotPetya. It needs to do that more often, and faster.

- **Rule Four:** We need to rethink the wisdom of reflexive secrecy around our cyber capabilities. Certainly, some secrecy about how our cyber weapons work is necessary— though by now, after Snowden and Shadow Brokers, there is not much mystery left. America's adversaries have a pretty complete picture of how the United States breaks into the darkest corners of cyberspace. We cannot expect Russian and Iranian hackers to stop implementing malware in our utility grid unless we are willing to talk about giving up our own implants in their power grids. We cannot insist that the U.S. government has the right to backdoor into Apple's iPhones and encrypted apps unless we are willing to make the Internet less safe for everyone, because any back door will become the target of hackers around the globe.

- **Rule Five:** The world needs to move ahead with setting these norms of behavior even if governments are not yet ready. Classic arms-control treaties won't work: they take years to negotiate and more to ratify. With the blistering pace of technological change in cyber, they would be outdated before they went into effect. The best hope is to reach a consensus on principles that begins with minimizing the danger to ordinary civilians, the fundamental political goal of most rules of warfare. A possible analogy is a [wartime] 'Digital Geneva Convention,' in which companies—not countries—take the lead in the short term.But companies must then step up their games too.[526]

Peter Singer and Alan Friedman agree with David Sanger that realistic solutions to cybersecurity threats are far from simple.[527]

A key problem to wrestle with is how to draw a strong distinction between attacks on military versus civilian facilities. Singer and Friedman offer, as an example that supposes that you are flying a bomber plane and are allowed to target an enemy's military vehicles. At the same time, your orders are to do your utmost to avoid hitting civilian vehicles, special vehicles like ambulances in particular.

Such a distinction isn't nearly so clear-cut in cyberspace, where a network can simultaneously be both civilian and military.

In all likelihood, a large state-sponsored cyberattack would intentionally attempt to do both. Singer and Friedman believe there is little chance that a nation-state would stage a major cyberattack against the U.S. without trying to cripple DoD in the process.

Why?

Because, as they point out, while a nation-state actor might try to cripple our country and our will by destroying private-sector systems like the power grid, pipelines, transportation, or banking, it is hard to imagine such actions come as a bolt from the blue:

> *Cyberattacks would only likely come in a period of heightened tensions between the U.S. and the attacker nation. In such an atmosphere, the attacker would probably already fear the possibility of conventional, or kinetic, action by the U.S. military.*
>
> *Moreover, if an opponent were going to hit U.S. with a large cyber attack, they would have to assume that we might respond kinetically. A cyberattack on the U.S. military would likely concentrate on DoD's networks.*[528]

All three types of DoD networks have penetration and disruption vulnerabilities.

Although NIPRNET is unclassified intranet…systems on the network use dot-mil addresses which connect to the Internet, this doesn't mean that it isn't important. It supports logistical functions such as supplying Army units with food. Such information can provide an enemy with many strategic data about troop locations, numbers, etc.

SIPRNET, which is supposed to be air-gapped between unclassified and secret-level networks, has been breached by users who inadvertently download malware from the Internet and pass it along to the classified side. Pentagon information security specialists refer to this problem as a "sneakernet threat."

Pentagon SIPRNET computers were successfully targeted for just such a sneakernet attack in November 2008 when a Russian-origin piece of spyware circulated cyberspace searching for vulnerable dot-mail addresses, hacked its way into the unclassified NIPRNET computers, began liking for thumb drives, and then downloaded itself into them.

Within hours, the spyware had infected thousands of secret-level U.S. military computers in Afghanistan, Iraq, Qatar, and elsewhere in the Central Command.

There are more than 100,000 SIPRNET terminals around the world. If you can get time alone with one terminal for a few minutes, you can upload malware or run a covert connection to the Internet.

In short, as Singer and Friedman observed, "computers on DoD's most important network had less protection than you probably have on your home computer." [529]

DoD's highest security-level JWIS network terminals are located in highly secured rooms known as Secret Compartmentalized Information Facilities (SCIFs) which are often referred to as "the Vaults". Nevertheless, although access is limited, information flowing on the network still has to go across fiber optic cables and through routers and servers, just as with any other network.

JWIS routers can be attacked to cut communications. In addition, the hardware used in the network's computers, servers, routers, and switches can potentially all be compromised at the point of manufacture or later on.

Encouragingly, Singer and Friedman report that the DoD has embarked on an extensive program to upgrade the security on all three kinds of these networks. Yet even if its networks are secure, DoD still runs risks that the software and/or hardware it has running its weapons systems may be compromised.

Just as plans for the new F-35 fighter were stolen by hacking into a defense contractor, hypothetical potentials also exist for a hacker to insert a hidden logic bomb within the millions of lines of code, or in the numerous pieces of firmware and computer hardware that run the aircraft. Such malware might, for example, cause the aircraft—or perhaps a missile system—to malfunction in the air upon receiving a certain radio command transmitted via an enemy drone or air defense station.

Singer and Friedman urge that one of the most important things the Pentagon can do is to develop a rigorous engineering design, inspection, and research program to ensure that the software and hardware being used in a key weapons system, in command control, and in logistics are not laced with malware.

The CIA has recently set forth what it terms a "Cyber Security Triad" of guiding principles that apply equally to all private, commercial and government/military networks.

The first, "confidentiality", entails keeping sensitive information private. It emphasizes the application of encryption services to protect data at rest or in transit to prevent unauthorized access by bad actors.

"Integrity" addresses the protection of data, networks, and systems from outside tampering. This includes mitigation and proactive measures to restrict unapproved changes, while also retaining the ability to recover data that has been lost or compromised.

"Availability" refers to ensuring that authorized users can freely access the systems, networks and data needed to perform their daily tasks. This includes means to resolve hardware and software conflicts as well as to provide critical maintenance to keep systems up and available.

Vital national security priorities underpinning all of the CIA Triad principles are to harden the Internet backbone, separate and secure the controls for the power grid and other critical utilities, and to vigorously pursue security upgrades for DoD IT systems which crucially defend all other networks.

As a 2018 DoD strategic cybersecurity report reaffirms, "cyberspace is critical to the way the entire U.S. functions." The document also confirms that White House defense policies "allow the military to gain an informational advantage, [and to] strike targets remotely and work from anywhere in the world."[530]

The DoD's cyberdefense mandate is not limited to nation-state military actors, but more broadly includes terrorists, criminals, and foreign adversaries such as Russia and China who are using cyber to try to steal our technology, disrupt our economy and government processes, and threaten critical infrastructure.

The DoD strategy clarifies a preemptive need "to preserve U.S. cyberspace superiority and stop cyberattacks before they hit our networks" through the application of five key action pillars:

One: Build a More Lethal Force

Troops have to increasingly worry about cyberattacks while still achieving their missions, so the DoD needs to make processes more flexible. Here's how:

Capabilities are going to be more diverse and adaptable.

Automation and large-scale data analytics will help identify cyberattacks and make sure our systems are still effective.

More commercial technology will be integrated

into current systems for maximum effectiveness in the ever-changing cybersphere.

Two: Compete and Deter in Cyberspace

This means preventing harmful cyber activities before they happen by:

Strengthening the cybersecurity of systems and networks that support DoD missions, including those in the private sector and our foreign allies and partners.

Streamlining public-private information-sharing.

Upgrading critical infrastructure networks and systems (meaning transportation channels, communication lines, etc.) to reduce risks of major cyberattacks on them.

Setting and enforcing standards for cybersecurity, resilience and reporting.

Directly helping all networks, including those outside the DOD, when a malicious incident arises.

Three: Strengthen Alliances and Attract New Partnerships

We can't do this mission alone, so the DoD must expand its cyber cooperation by:

Building dependable partnerships with private-sector entities who are vital to helping support military operations.

Sharing information with other federal agencies, our agencies, and foreign partners and allies who have advanced cyber capabilities. This will increase effectiveness.

Looking for crowdsourcing opportunities such as hack-a-thons and bug bounties to identify and fix our vulnerabilities.

Upholding cyberspace behavioral norms during peacetime.

Four: Reform the Department

Personnel must increase their cyber awareness. The DoD is making strides in this by:

Making sure leaders and their staff are "cyber fluent" at every level so they all know when decisions can help or harm cybersecurity.

Holding DoD personnel and third-party contractors more accountable for slip-ups.

Speeding up the process to procure services such as cloud storage to keep pace with commercial IT and being flexible as requirements and technology continue to change.

Five: Cultivate Talent

Retaining the current cyber workforce is key, as is finding talented new people to recruit. The department will do this by:

Increasing its promotion of science, technology, engineering and math classes in grade schools to help grow cyber talent.

Creating competitions and other processes to identify top-tier cyber specialists who can help with DoD's toughest challenges.

Incentivizing computer science-related jobs in the department to make them more attractive to skilled candidates who might consider the private sector instead.

Examples are rotational billets for service members at other federal agencies; specialized training opportunities; expansion of compensation incentives; and optimizing the mix of service members, civilians and contractors who can best support the mission.[531]

Trump White House and DoD cybersecurity program mandates reflect urgently overdue reversal of previous "lead from behind" policies. Nevertheless, government efforts alone won't be sufficient to make U.S. safe.

David Sanger urges U.S. to recognize that a major obstacle to achieving such an ambitious goal is that so many of the most vulnerable key targets are in private hands. Due to Internet complexity, the government can't regulate how banks, telecom firms, gas pipeline companies, Google and Facebook design their cybersecurity. Every one of these systems is radically different.[532]

Sanger argues that there is a prevalently mistaken notion that since the Department of Homeland Security is supposed to "coordinate" with the Pentagon to defend the United States against incoming missile attacks, that by extension, it will defend American companies and individuals against sophisticated, state-sponsored hacks (but excluding scammers, teenage hackers, and trolls living in Saint Petersburg).

Sanger argues that it's time to *get real* on such matters because the government will play no role in protecting American institutions except where it comes to the most critical of infrastructures: the electric grid, the voting system, the water and wastewater systems, the financial system, and nuclear weapons.

At best, Sanger predicts, predicts, "We would be lucky to seal up three-quarters of the glaring vulnerabilities in American networks today." Unless we adjust our strategy to reflect this reality, "we will be far more vulnerable than almost any other major nation for years to come."

Sanger believes that the best way to deter attacks—and also to counterattack—is deterrence by denial of cyber-secure networks.

"Getting real" to accomplish this will require a major national effort, one far beyond the civil defense projects of the 1950s, when the United States built a highway system that could evacuate civilians and dug shelters in large cities.[533]

Enormously ambitious and costly? Without doubt.

Meanwhile, China's all-out economic commitment to challenge America's security in cyberspace is beyond question. Chinese government hackers are known to have penetrated thousands of U.S. communications networks and to have inserted prospectively destructive malware into America's power grid.

Beijing's goal of world supremacy in quantum computing must be recognized as a particularly clear and rapidly escalating threat.

Russia, already a nuclear superpower, is now actively turning to far less expensive cyberwarfare DDoS campaigns of military and economic conquest fronted for deniability by armies of "patriot" cybercriminals. Cyberattacks on Ukraine power transfer and Internet communication networks should be regarded as important U.S. lessons.

As with China, Russian operatives have already penetrated the control systems of some U.S. electric power companies.

Economically backward North Korea is cultivating and training a new generation of highly sophisticated hackers. Previous attacks on the Sony Corporation, the Bangladesh Central Bank, and even U.S. and South Korean government websites are but sample harbingers of far the greater damage that even a small hermit rogue nation can wreak upon economically and technologically advanced powers.

Whether or not their nuclear-tipped ICBMs can reach America's mainland, their weaponized malware has no geographical limitations whatsoever.

Cyberattacks that froze financial networks of the Bank of America and Chase, fried computers at the Sands Casino also demonstrate Iran's appetite and emerging capacity for the ever-greater U.S. and global intimidation leverage.

Again, let's all be reminded of the grim prospective consequences should America fail to maintain decisive offensive and defensive cyberwarfare leadership.

A major cyberattack on America, for example, could cause every bit as much—and even more—social and economic devastation as would a nuclear electromagnetic pulse (EMP) on our electric power grid.

First' it's terrifying enough trying to contemplate the chaos and destruction that would accompany a North Korean-launched EMP detonation from orbit over Chicago, for example. Dare to imagine circumstances with grid disruptions shutting down all water pumping and sanitation stations, lights, refrigerators and air-conditioning, TV, radio and Internet communications; air, and; surface transportation fuel and food delivery, law enforcement and emergency rescue operations; and hospital critical care services; and manufacturing industries.

Now further, try to contemplate additional disruptions of a major cyber-attack upon the energy grid along with all other networks and equipment connected to the vast Internet of Things. Vulnerable cyber targets will predictably include federal, state and local government agencies; military defense systems and forces; and international banking and business networks.

Are such unimaginable scenarios truly possible after all?

America can ill-afford the incomprehensible costs of assuming and discovering otherwise.

[1] *Cyber War*, Richard Clarke and Robert K. Knake, 2010, HarperCollins.

[2] *Lights Out: A Cyberattack, A Nation Unprepared, Surviving the Aftermath*, Ted Koppel, 2015, Broadway Books, an imprint of the Crown Publishing Group, a division of penguin Random House, LLC.

[3] *The Perfect Weapon,* David E. Sanger, 2018, Crown Publishing Group, Penguin Random House LLC, New York.

[4] Ibid.

[5] *The Fifth Domain: Defending Our Country, Our Companies, and Ourselves in the Age of Cyber Threats*, Richard A Clarke and Robert K. Knake, 2019, New York, Penguin Press.

[6] *The Perfect Weapon*, David E. Sanger, 2018, Crown Publishing Group, Penguin Random House LLC, New York.

[7] Ibid.

[8] Ibid.

[9] Ibid.

[10] *Cyber War,* Richard Clarke and Robert K. Knake, 2010, HarperCollins.

[11] Ibid.

[12] Ibid.

[13] *FERC Requires Expanded Cybersecurity Incident Reporting,* Federal Energy Regulatory Commission, July 19, 2018, www.ferc.gov/media/news-releases/2018/2018-3/07-19-18-E-1.asp.

[14] The Fifth Domain: Defending Our Country, Companies, and Ourselves in the Age of Cyber Threats, Richard A. Clarke and Robert K Knake, 2019, Penguin Press.

[15] Ibid.

[16] Ibid.

[17] *Electricity Grid in U.S. Penetrated By Spies,* Siobhan Gorman, April 8, 2009, Wall Street Journal.

[18] *New cyber attack hits Israeli stock exchange and airline*, Yolande Knell, January 16, 2012, BBC News, Jerusalem.

[19] *Cyber War,* Richard Clarke and Robert K. Knake, 2010, HarperCollins.

[20] *The Fifth Domain: Defending Our Country, Our Companies, and Ourselves in the Age of Cyber Threats*, Richard A Clarke and Robert K. Knake, 2019, New York, Penguin Press.

[21] Ibid.

[22] Ibid.

[23] *Cybersecurity and Cyberwar: What Everyone Needs to Know,* P.W. Singer and Alan Friedman, 2014, Oxford University Press .

[24] Ibid.

[25] *Cyber War,* Richard Clarke and Robert K. Knake, 2010, HarperCollins.

[26] *Cybersecurity and Cyberwar: What Everyone Needs to Know,* P.W. Singer and Alan Friedman, 2014, Oxford University Press.

[27] Ibid.

[28] Ibid.

[29] Ibid.

[30] Ibid.

[31] *Israeli Test on Worm Called Crucial in Iran Nuclear Delay,* William J. Broad, John Markoff, and David E. Sanger, January 15, 2011, The New York Times.

[32] *The Real Story of Stuxnet,* D. Kushner, IEEE Spectrum 53, No. 3, 48 (2013).

[33] *Cybersecurity and Cyberwar: What Everyone Needs to Know,* P.W. Singer and Allan Friedman, 2014, Oxford University Press.

[34] Ibid.

[35] *The Perfect Weapon,* David E. Sanger, 2018, Crown Publishing Group, Penguin Random House LLC, New York.

[36] Ibid.

[37] Ibid.

[38] Ibid.

[39] *Cybersecurity and Cyberwar: What Everyone Needs to Know,* P.W. Singer and Allan Friedman, 2014, Oxford University Press.

[40] *The Perfect Weapon,* David E. Sanger, 2018, Crown Publishing Group, Penguin Random House LLC, New York.

[41] *Cybersecurity and Cyberwar: What Everyone Needs to Know,* P.W. Singer and Allan Friedman, 2014, Oxford University Press.

[42] *The Perfect Weapon,* David E. Sanger, 2018, Crown Publishing Group, Penguin Random House LLC, New York.

[43] *Cybersecurity and Cyberwar: What Everyone Needs to Know,* P.W. Singer and Allan Friedman, 2014, Oxford University Press.

[44] Ibid.

[45] *Cyber War,* Richard Clarke and Robert K. Knake, 2010, HarperCollins.

[46] *What is cyberwar? Everything you need to know about the frightening future of digital conflict,* Steve Ranger, December 4, 2018, ZDNet Week in Review—the US.

[47] *The Perfect Weapon,* David E. Sanger, 2018, Crown Publishing Group, Penguin Random House LLC, New York.

[48] *Cyber War,* Richard Clarke and Robert K. Knake, 2010, HarperCollins.

[49] *Cybersecurity and Cyberwar: What Everyone Needs to Know,* P.W. Singer and Allan Friedman, 2014, Oxford University Press.

[50] Ibid.

[51] *Cyber War,* Richard Clarke and Robert K. Knake, 2010, HarperCollins.

[52] Ibid.

[53] *The Perfect Weapon,* David E. Sanger, 2018, Crown Publishing Group, Penguin Random House LLC, New York.

[54] Ibid.

[55] Ibid.

[56] *Cybersecurity and Cyberwar: What Everyone Needs to Know,* P.W. Singer and Alan Friedman, 2014, Oxford University Press.

[57] *Cyber War,* Richard Clarke and Robert K. Knake, 2010, HarperCollins.

[58] Ibid.

[59] *Cybersecurity and Cyberwar: What Everyone Needs to Know,* P.W. Singer and Alan Friedman, 2014, Oxford University Press.

[60] *Cyber War,* Richard Clarke and Robert K. Knake, 2010, HarperCollins.

[61] *The Israeli 'E-tack' on Syria—Part I,* March 9, 2008, U.S. Air Force Technology.

[62] *Cyber War,* Richard Clarke and Robert K. Knake, 2010, HarperCollins.

[63] Ibid.

[64] *The Fifth Domain: Defending Our Country, Our Companies, and Ourselves in the Age of Cyber Threats,* Richard A Clarke and Robert K. Knake, 2019, New York, Penguin Press.

[65] *Cyber War,* Richard Clarke and Robert K. Knake, 2010, HarperCollins.

[66] *The Perfect Weapon,* David E. Sanger, 2018, Crown Publishing Group, Penguin Random House LLC, New York.

[67] Ibid.

[68] *The First Cyber Espionage Attacks: How Operation Moonlight Maze made history,* Chris Doman, July 7, 2016, Medium.com.

[69] Ibid.

[70] *The Perfect Weapon,* David E. Sanger, 2018, Crown Publishing Group,

Penguin Random House LLC, New York.

[71] Ibid.

[72] *Our Country, Our Companies, and Ourselves in the Age of Cyber Threats*, Richard A Clarke and Robert K. Knake, 2019, New York, Penguin Press.

[73] *The Untold Story of NotPetya, the Most Devastating Cyberattack in History*, Andy Greenberg, August 22, 2018, Wired.com .

[74] *The Fifth Domain: Defending Our Country, Our Companies, and Ourselves in the Age of Cyber Threats*, Richard A Clarke and Robert K. Knake, 2019, New York, Penguin Press.

[75] Ibid.

[76] Ibid.

[77] Ibid.

[78] *Russian operation hacked a Vermont utility, showing risk to U.S. electrical grid security, officials say*, Juliet Eilperin and Adam Entous, December 31, 2016, Washington Post-National Security.

[79] *The Fifth Domain: Defending Our Country, Our Companies, and Ourselves in the Age of Cyber Threats*, Richard A Clarke and Robert K. Knake, 2019, New York, Penguin Press.

[80] Ibid.

[81] *The Perfect Weapon*, David E. Sanger, 2018, Crown Publishing Group, Penguin Random House LLC, New York.

[82] Ibid.

[83] Ibid.

[84] *Cybersecurity and Cyberwar: What Everyone Needs to Know*, P.W. Singer and Alan Friedman, 2014, Oxford University Press.

[85] *The Fifth Domain: Defending Our Country, Our Companies, and Ourselves in the Age of Cyber Threats*, Richard A Clarke and Robert K. Knake, 2019, New York, Penguin Press.

[86] *Cyber War*, Richard Clarke and Robert K. Knake, 2010, HarperCollins.

[87] *Cybersecurity and Cyberwar: What Everyone Needs to Know*, P.W. Singer and Alan Friedman, 2014, Oxford University Press.

[88] Ibid.

[89] Ibid.

[90] Ibid.

[91] Ibid.

[92] Ibid.

[93] *Cyber War,* Richard Clarke and Robert K. Knake, 2010, HarperCollins.

[94] Ibid.

[95] Ibid.

[96] Ibid.

[97] *Cybersecurity and Cyberwar: What Everyone Needs t Know,* P.W. Singer and Alan Friedman, 2014, Oxford University Press.

[98] *Cyber War,* Richard Clarke and Robert K. Knake, 2010, HarperCollins.

[99] Ibid.

[100] Ibid.

[101] Ibid.

[102] *Deceiving the Sky: Inside Communist China's Drive for Global Supremacy,* Bill Gertz, 2019, Encounter Books.

[103] *Cyber War,* Richard Clarke and Robert K. Knake, 2010, HarperCollins.

[104] *Deceiving the Sky: Inside Communist China's Drive for Global Supremacy,* Bill Gertz, 2019, Encounter Books.

[105] Ibid.

[106] *China's Tech Champion—Or Serial Thief?,* Chuin-Wei Yap and Dan Strumpf (Hong Kong) and Dustin Volz, Kate O'Keefe and Aruna Viswanatha (Washington, DC), May 25-26, 2019, Wall Street Journal.

[107] *Deceiving the Sky: Inside Communist China's Drive for Global Supremacy,* Bill Gertz, 2019, Encounter Books.

[108] Ibid.

[109] Ibid.

[110] Ibid.

[111] Ibid.

[112] Ibid.

[113] Ibid.

[114] Ibid.

[115] Ibid.

[116] Ibid.

[117] Ibid.

[118] Ibid.

[119] Ibid.

[120] Ibid.

[121] *Cyber War,* Richard Clarke and Robert K. Knake, 2010, HarperCollins.

[122] *China's Tech Champion—Or Serial Thief?,* Chuin-Wei Yap and Dan

Strumpf (Hong Kong) and Dustin Volz, Kate O'Keefe and Aruna Viswanatha (Washington, DC), May 25-26, 2019, Wall Street Journal .

[123] Ibid.

[124] *Guilty verdict in the theft of Motorola secrets for China*, February 8, 2012, Phys.org.

[125] *Huawei's Yearslong Rise Is Littered With Accusations of Theft and Dubious Ethics,* Chuin-Wei Yap, Dan Strumpf, Dustin Volz, Kate O'Keefe, and Aruna Viswanatha, May 29, 2019, Wall Street Journal.

[126] *The Perfect Weapon,* David E. Sanger, 2018, Crown Publishing Group, Penguin Random House LLC, New York.

[127] *What is cyberwar? Everything you need to know about the frightening future of digital conflict*, Steve Ranger, December 4, 2018, ZDNet Week in Review—the US.

[128] *The Perfect Weapon,* David E. Sanger, 2018, Crown Publishing Group, Penguin Random House LLC, New York.

[129] *Ibid.*

[130] *The U.S. to Expand Huawei Restrictions,* Catherine Lucey and Dan Strumpf, August 8, 2019, the Wall Street Journal.

[131] Ibid.

[132] *Cyber War,* Richard Clarke and Robert K. Knake, 2010, HarperCollins.

[133] Ibid.

[134] Ibid.

[135] Ibid.

[136] Ibid.

[137] Ibid.

[138] Ibid.

[139] Ibid.

[140] Ibid.

[141] Ibid.

[142] Ibid.

[143] Ibid.

[144] *North Korea Reaps Cash, Leverage From Cyber Skills,* Ian Talley and Dustin Volz, September 16, 2019, Wall Street Journal.

[145] Ibid.

[146] *The Perfect Weapon,* David E. Sanger, 2018, Crown Publishing Group, Penguin Random House LLC, New York.

[147] Ibid.

[148] *North Korea Reaps Cash, Leverage From Cyber Skills,* Ian Talley and Dustin Volz, September 16, 2019, Wall Street Journal.

[149] Ibid.

[150] *Are we underestimating Iran's cyber capabilities?,* Annie Fixler, March 11, 2019, The Hill.

[151] Ibid.

[152] *The Perfect Weapon,* David E. Sanger, 2018, Crown Publishing Group, Penguin Random House LLC, New York.

[153] Ibid.

[154] *Are we underestimating Iran's cyber capabilities?,* Annie Fixler, March 11, 2019, The Hill.

[155] *Cybersecurity Firms Warn if New Risks: Current and former U.S. Officials Say Tehran May Retaliate with Hacking Attacks,* Dustin Volz, June 22-23, 2019, Wall Street Journal.

[156] Ibid.

[157] *Are we underestimating Iran's cyber capabilities?,* Annie Fixler, March 11, 2019, The Hill.

[158] Ibid.

[159] *Suspected Iranian Hackers Hit Bahrain,* Bradley Hope, Warren P. Stobel and Dustin Volz, August 8, 2019, the Wall Street Journal.

[160] Ibid.

[161] Ibid.

[162] *Cybersecurity Firms Warn if New Risks: Current and former U.S. Officials Say Tehran May Retaliate with Hacking Attacks,* Dustin Volz, June 22-23, 2019, Wall Street Journal.

[163] *The Perfect Weapon,* David E. Sanger, 2018, Crown Publishing Group, Penguin Random House LLC, New York.

[164] *Iran Blamed for Cyberattacks,* Siobhan Gorman and Julian E. Barnes, October 12, 2012, Wall Street Journal.

[165] *The Fifth Domain: Defending Our Country, Our Companies, and Ourselves in the Age of Cyber Threats,* Richard A Clarke and Robert K. Knake, 2019, New York, Penguin Press.

[166] *The Perfect Weapon,* David E. Sanger, 2018, Crown Publishing Group, Penguin Random House LLC, New York.

[167] *Don't Look, Don't Read: Government Warns Its Workers Away From WikiLeaks Documents,* Eric Lipton, December 4, 2010, New York

Times.com.

[168] *Manning-Lamo Chat Logs Revealed,* Evan Hansen, Threat Level (blog), Wired, July 13, 2011.

[169] *Pentagon Sees a Threat from Online Muckrakers,* Stephanie Strom, March 17, 2010, New York Times..

[170] *Manning-Lamo Chat Logs Revealed,* Evan Hansen, Threat Level (blog), Wired, July 13, 2011.

[171] *The Perfect Weapon,* David E. Sanger, 2018, Crown Publishing Group, Penguin Random House LLC, New York.

[172] Ibid.

[173] Ibid.

[174] Ibid.

[175] Ibid.

[176] Ibid.

[177] Ibid.

[178] Ibid.

[179] *The Fifth Domain: Defending Our Country, Our Companies, and Ourselves in the Age of Cyber Threats*, Richard A Clarke and Robert K. Knake, 2019, New York, Penguin Press.

[180] *Vault 7: CIA Hacking Tools Revealed*, https://wikileaks.org/ciav7p1/.

[181] Ibid.

[182] Ibid.

[183] *The Fifth Domain: Defending Our Country, Our Companies, and Ourselves in the Age of Cyber Threats*, Richard A Clarke and Robert K. Knake, 2019, New York, Penguin Press.

[184] *Vault 7: CIA Hacking Tools Revealed*, https://wikileaks.org/ciav7p1/.

[185] *Capital One Breach Affects 106 Million Applicants,* Nicole Hong, Liz Hoffman, and AnnaMaria Andriotos, the Wall Street Journal, July 30, 2019.

[186] *Hacking Suspect Left Trail of Clues Online,* Dana Mattioli, Robert McMillan, and Sebastian Herrera, July 31, 2019, the Wall Street Journal.

[187] *Hack Casts a Shadow on Cloud Security,* Robert McMillan, July 31, 2019, the Wall Street Journal.

[188] Ibid.

[189] *Capital One's Data Breach: What It Means for You,* Bourree Lam and Julia Carpenter, July 31, 2019, the Wall Street Journal.

[190] Ibid.

[191] Ibid.

[192] *The Perfect Weapon,* David E. Sanger, 2018, Crown Publishing Group, Penguin Random House LLC, New York.

[193] Ibid.

[194] Ibid.

[195] Ibid.

[196] Ibid.

[197] Ibid.

[198] *Cyber War,* Richard Clarke and Robert K. Knake, 2010, HarperCollins.

[199] *Cyber Warfare: The New Front,* Marie O'Neill Sciarrone, George W. Bush Institute, Spring, 2017.

[200] *The Weaponization of AI and the Internet: How Global Networks of Infotech Overlords Are Expanding Their Control Over Our Lives*, Larry Bell, 2019, Stairway Press.

[201] *U.S. Department of Homeland Security Strategy,* May 15, 2018, https://www.dhs.gov/sites/default/files/publications/DHS-Cybersecurity-Strategy_1.pdf.

[202] *China's 'National Security' Hypocrisy,* Jeff Moon, June 16, 2019, Wall Street Journal.

[203] *The Perfect Weapon,* David E. Sanger, 2018, Crown Publishing Group, Penguin Random House LLC, New York.

[204] Ibid.

[205] Ibid.

[206] Ibid.

[207] *Cyber War,* Richard Clarke and Robert K. Knake, 2010, HarperCollins.

[208] *The Perfect Weapon,* David E. Sanger, 2018, Crown Publishing Group, Penguin Random House LLC, New York.

[209] Ibid.

[210] *Cyber War,* Richard Clarke and Robert K. Knake, 2010, HarperCollins.

[211] *Cybersecurity and Cyberwar: What Everyone Needs to Know,* P.W. Singer and Alan Friedman, 2014, Oxford University Press.

[212] *China is eroding the U.S. edge in AI and 5G,* Kaveh Waddedell and Erica Pandey, September 25, 2019, AXIOS.

[213] *China's 'National Security' Hypocrisy,* Jeff Moon, June 16, 2019, Wall Street Journal.

[214] *The Autocrat's New Tool Kit,* Richard Fountaine and Kara Frederick,

March 15, 2019, Wall Street Journal.

[215] *Chinese Media Claims NYPD Is Using Beijing-Controlled Facial Recognition—Is It True?*, Zak Doffman, Forbes.com, January 13, 2019.

[216] *The Autocrat's New Tool Kit,* Richard Fountaine and Kara Frederick, March 15, 2019, Wall Street Journal.

[217] *From Falun Gong to Xinjiang: China's Repression Maestro*, Chun Han Wong, April 4, 2019, Wall Street Journal.

[218] Ibid.

[219] " Ibid.

[220] *China selling high-tech tyranny to Latin America, stoking U.S. concern*, Joel Gehrke, April 10, 2019, the Washington Examiner.

[221] Ibid.

[222] *The Autocrat's New Tool Kit,* Richard Fontaine and Kara Frederick, March 16-17, Wall Street Journal Review Section.

[223] *The Pros and Cons of Facial Recognition,* Ted Rall, May 16, 2019, The Wall Street Journal.

[224] *The Chinese technology helping New York police keep a closer eye on the United States' biggest city*, Stephen Chen, January 11, 2019, South China Post.

[225] Ibid.

[226] Ibid.

[227] Ibid.

[228] *Gifts That Snoop? The Internet of Things Is Wrapped in Privacy Concerns*, Bree Fowler, December 13, 2017, Consumer Reports.

[229] *FTC WARNS OF SECURITY AND PRIVACY RISKS IN IOT DEVICES*, Pindrop.com.

[230] *The Autocrat's New Tool Kit,* Richard Fontaine and Kara Frederick, March 16-17, Wall Street Journal Review Section.

[231] *FTC WARNS OF SECURITY AND PRIVACY RISKS IN IOT DEVICES*, Pindrop.com.

[232] *The Weaponization of AI and the Internet; How Global Networks of Infotech Overlords Are Expanding their Control Over Our Lives*, Larry Bell, 2019, Stairway Press.

[233] *The truth about smart cities: 'In the end, they will destroy democracy*, Steven Poole, December 17, 2014, The Guardian.

[234] *The smart entrepreneurial city: Dholera and 100 other utopias in India,*

S. Marvin, A. Luque-Ayala, and C McFarlane (Eds.), 2015, Smart urbanism: Utopian vision or false dawn? Rutledge.

[235] *The truth about smart cities: 'In the end, they will destroy democracy,* Steven Poole, December 17, 2014, The Guardian.

[236] Ibid.

[237] *The Autocrat's New Tool Kit,* Richard Fontaine and Kara Frederick, March 16-17, 2019, Wall Street Journal Review Section.

[238] "Ibid.

[239] Ibid.

[240] Ibid.

[241] Ibid.

[242] Ibid.

[243] Ibid.

[244] Ibid.

[245] *Facebook artificial intelligence spots suicidal users,* Leo Kelion, March 1, 2017, BBC News.

[246] *Deep Learning 'Godfather' Bengio Worries About China's Use of AI,* Jeremy Kahn, February 2, 2019, Bloomberg.

[247] *The Weaponization of AI and the Internet: How Global Networks of Infotech Overlords Are Expanding Their Control Over Our Lives,* Larry Bell, 2019, Stairway Press.

[248] *The Autocrat's New Tool Kit,* Richard Fontaine and Kara Frederick, March 16-17, 2019, Wall Street Journal Review Section.

[249] *Cybersecurity and Cyberwar: What Everyone Needs to Know,* P.W. Singer and Allan Friedman, 2014, Oxford University Press.

[250] Ibid.

[251] Ibid.

[252] Ibid.

[253] Ibid.

[254] Ibid.

[255] Ibid.

[256] Ibid.

[257] *The Perfect Weapon,* David E. Sanger, 2018, Crown Publishing Group, Penguin Random House LLC, New York.

[258] Ibid.

[259] Ibid.

[260] *What the FBI Files Reveal About Hillary Clinton's Email Server,*

Garrett. M. Graff, September 30, 2016, Politico.

[261] *What Russia's DNC Hack Tells U.S. About Hillary Clinton's Private Email Server*, Paul Roderick Gregory, June 15, 2016, Forbes.com .

[262] *The Perfect Weapon,* David E. Sanger, 2018, Crown Publishing Group, Penguin Random House LLC, New York.

[263] Ibid.

[264] Ibid.

[265] Ibid.

[266] Ibid.

[267] Ibid.

[268] Ibid.

[269] Ibid.

[270] Ibid.

[271] Ibid.

[272] *Mueller Report: Special Counsel Didn't Examine DNC Servers—Based on FBI Investigation that Didn't Examine DNC Servers*, Aaron Klein, May 1, 2019, Breitbart.

[273] Ibid.

[274] Ibid.

[275] Ibid.

[276] Ibid.

[277] *The Perfect Weapon,* David E. Sanger, 2018, Crown Publishing Group, Penguin Random House LLC, New York.

[278] *Russia Isn't the Only One Meddling in Elections. We Do It, Too*, Scott Shane, February 17, 2018, The New York Times.

[279] Ibid.

[280] Ibid.

[281] Ibid.

[282] *The Perfect Weapon,* David E. Sanger, 2018, Crown Publishing Group, Penguin Random House LLC, New York.

[283] Ibid.

[284] Ibid.

[285] Ibid.

[286] Ibid.

[287] Ibid.

[288] Ibid.

[289] Ibid.

[290] Ibid.

[291] Ibid.

[292] Ibid.

[293] Ibid.

[294] Ibid.

[295] *What Is the Dark Web—Who Uses It, Dangers & Precautions to Take*, Michael Lewis, May 2019, Moneycrashers.

[296] Ibid.

[297] *Deeplight: Shining the Light On the Dark Web*, An Intelliagg Report, 2016.

[298] *What Is the Dark Web—Who Uses It, Dangers & Precautions to Take*, Michael Lewis, May 2019, Moneycrashers.

[299] *The most Dangerous Hackers Today,* Joseph Regan, July 24, 2018, AVG.com.

[300] Ibid.

[301] *It's not just Russia. China, North Korea, and Iran could interfere in 2018 elections too*, Alex Ward, August 20, 2018, Vox .

[302] *The most Dangerous Hackers Today*, Joseph Regan, July 24, 2018, AVG.com.

[303] *New Chinese Hacker Group Linked to 'One Belt, One Road' Initiative*, Frank Fang, March 7-13, The Epoch Times.

[304] *The most Dangerous Hackers Today*, Joseph Regan, July 24, 2018, AVG.com.

[305] Ibid.

[306] *North Korea hackers are targeting 'critical' U.S. infrastructure, a cybersecurity firm says*, Donie O'Sullivan and Joshua Berlinger, March 5, 2019, CNN.

[307] *The most Dangerous Hackers Today*, Joseph Regan, July 24, 2018, AVG.com.

[308] *Are we underestimating Iran's cyber capabilities?*, Annie Fixler, March 11, 2919, The Hill.

[309] *2 Iranian men indicted for city of Atlanta ransomware attack*, November 28, 2018, Fox 5 Atlanta.

[310] Ibid.

[311] *Confidential Report: Atlanta's cyber attack could cost taxpayers $17 million*, Stephen Deere, August 1, 2018, The Atlanta Journal-Constitution.

[312] Ibid.

[313] Ibid.

[314] *Cybersecurity and Cyberwar,* P.W. Singer and Allan Friedman, 2014, Oxford University Press.

[315] *The Perfect Weapon,* David E. Sanger, 2018, Crown Publishing Group, Penguin Random House LLC, New York.

[316] *Cybersecurity and Cyberwar,* P.W. Singer and Allan Friedman, 2014, Oxford University Press.

[317] Ibid.

[318] Ibid.

[319] Ibid.

[320] Ibid.

[321] *Cyber War,* Richard Clarke and Robert K. Knake, 2010, HarperCollins.

[322] *Cybersecurity and Cyberwar: What Everyone Needs to Know,* P.W. Singer and Allan Friedman, 2014, Oxford University Press.

[323] Ibid.

[324] Ibid.

[325] Ibid.

[326] Ibid.

[327] *Drone terrorism is now a reality, and we need a plan to counter the threat,* Colin P. Clarke, August 20, 2018, Rand Corporation.

[328] Ibid.

[329] *Cybersecurity and Cyberwar: What Everyone Needs to Know,* P.W. Singer and Allan Friedman, 2014, Oxford University Press.

[330] *Al-Qaeda plans cyberattacks on dams,* Toby Harnden, June 28, 2002, The Telegraph.

[331] *Cybersecurity and Cyberwar: What Everyone Needs to Know,* P.W. Singer and Allan Friedman, 2014, Oxford University Press.

[332] *The Perfect Weapon,* David E. Sanger, 2018, Crown Publishing Group, Penguin Random House LLC, New York.

[333] *Cyber War,* Richard Clarke and Robert K. Knake, 2010, HarperCollins.

[334] Ibid.

[335] Ibid.

[336] Ibid.

[337] Ibid.

[338] *The Perfect Weapon,* David E. Sanger, 2018, Crown Publishing Group,

Penguin Random House LLC, New York.

[339] Ibid.

[340] *Cyber War,* Richard Clarke and Robert K. Knake, 2010, HarperCollins.

[341] Ibid.

[342] *The Fifth Domain,* Richard A. Clarke and Robert K. Knake, 2019, Penguin Press, New York.

[343] *Why 5G requires new approaches to cybersecurity; Racing to protect the most important network of the 21st century,* Tom Wheeler and David Simpson, September 3, 2019, Brookings Institute.

[344] Ibid.

[345] Ibid.

[346] *The Fifth Domain,* Richard A. Clarke and Robert K. Knake, 2019, Penguin Press, New York.

[347] *Deceiving the Sky: Inside Communist China's Drive for Global Supremacy,* Bill Gertz, 2019, Encounter Books, New York-London.

[348] Ibid.

[349] *EU Warns of 5G Risks, Casting a Wary Eye on Huawei,* Anna Isaac and Parmy Olson, October 12-13, 2019, London-The Wall Street Journal.

[350] Ibid.

[351] *Beyond Huawei—5G and U.S. National Security,* Daniel Zhang, June 7, 2019, Georgetown Security Studies Review.

[352] *FCC Answers The Threat From Huawei,* Ajit Pai, October 29, 2019, the Wall Street Journal.

[353] Ibid.

[354] Ibid.

[355] *Beyond Huawei—5G and U.S. National Security,* Daniel Zhang, June 7, 2019, Georgetown Security Studies Review.

[356] Ibid.

[357] Ibid.

[358] *Why 5G requires new approaches to cybersecurity; Racing to protect the most important network of the 21st century,* Tom Wheeler and David Simpson, September 3, 2019, Brookings Report.

[359] Ibid.

[360] Ibid.

[361] Ibid.

[362] Ibid.

[363] *Beyond Huawei—5G and U.S. National Security,* Daniel Zhang, June 7,

2019, Georgetown Security Studies Review.

[364] *Why 5G requires new approaches to cybersecurity; Racing to protect the most important network of the 21st century*, Tom Wheeler and David Simpson, September 3, 2019, Brookings Report.

[365] *The Perfect Weapon,* David E. Sanger, 2018, Crown Publishing Group, Penguin Random House LLC, New York .

[366] Ibid.

[367] Ibid.

[368] Ibid.

[369] *A Close Look at the NSA's Most Powerful Internet Attack Tool,* Nicholas Weaver, March 13, 2014, Wired.com.

[370] *The Perfect Weapon,* David E. Sanger, 2018, Crown Publishing Group, Penguin Random House LLC, New York .

[371] *Cybersecurity and Cyberwar: What Everyone Needs to Know,* P.W. Singer and Alan Friedman, 2014, Oxford University Press.

[372] Ibid.

[373] Ibid.

[374] *Deceiving the Sky: Inside Communist China's Drive for Global Supremacy,* Bill Gertz, 2019, Encounter Books, New York-London.

[375] *Quantum computing: The new moonshot in the cyber space race,* Vinod Vaikuntanathan, August 23, 2019, HELPNETSECURITY.

[376] Ibid.

[377] America's Enigma problem with China: The threat of quantum computing, Morgan Wright, March 5, 2018, The Hill.

[378] Ibid.

[379] Ibid.

[380] *Quantum computing: The new moonshot in the cyber space race,* Vinod Vaikuntanathan, August 23, 2019, HELPNETSECURITY.

[381] *It's time to protect the blockchain from quantum-enabled hackers,* Kiran Bhagotra, December 18, 2018, WIRED.

[382] Ibid.

[383] *Quantum computing: The new moonshot in the cyber space race,* Vinod Vaikuntanathan, August 23, 2019, HELPNETSECURITY.

[384] *The Fifth Domain,* Richard A. Clarke and Robert K. Knake, 2019, Penguin Press, New York.

[385] Ibid.

[386] Ibid.
[387] *The Perfect Weapon,* David E. Sanger, 2018, Crown Publishing Group, Penguin Random House LLC, New York.
[388] Ibid.
[389] *The Fifth Domain,* Richard A. Clarke and Robert K. Knake, 2019, Penguin Press, New York.
[390] *Quantum computing: The new moonshot in the cyber space race,* Vinod Vaikuntanathan, August 23, 2019, HELPNETSECURITY.
[391] *Cybersecurity and Cyberwar: What Everyone Needs to Know,* P.W. Singer and Alan Friedman, 2014, Oxford University Press.
[392] Ibid.
[393] Ibid.
[394] *Cyber Warfare: The New Front,* Marie O'Neill Sciarrone, Spring 2017 Report, George W. Bush Institute.
[395] Ibid.
[396] Ibid.
[397] *Cyber War,* Richard Clarke and Robert K. Knake, 2010, HarperCollins.
[398] *What is cyberwar? Everything you need to know about the frightening future of digital conflict,* Steve Ranger, December 4, 2018, ZDNet Week in Review—the US.
[399] Ibid.
[400] Ibid.
[401] Ibid.
[402] *Pentagon says 'aware' of China Internet rerouting,* Jim Wolf, November 19, 2019, Reuters—Technology News.
[403] *The Fifth Domain: Defending Our Country, Our Companies, and Ourselves in the Age of Cyber Threats,* Richard A Clarke and Robert K. Knake, 2019, New York, Penguin Press.
[404] Ibid.
[405] *The Perfect Weapon,* David E. Sanger, 2018, Crown Publishing Group, Penguin Random House LLC, New York .
[406] *What is cyberwar? Everything you need to know about the frightening future of digital conflict,* Steve Ranger, December 4, 2018, ZDNet Week in Review—the US.
[407] *Cyber Warfare: The New Front,* Marie O'Neill Sciarrone, Spring 2017 Report, George W. Bush Institute.
[408] *Cyber Conflict and Geopolitics—the Cold War's New Front,* Victor

Hvozd, Diary of Economic Forum in Krynica Zdrój (Poland) themed "A Europe of Common Values or A Europe of Common Interests?", September 4-6, 2018, Borysfen Intel.

[409] Ibid.

[410] Ibid.

[411] *New Delhi: "Military to get a new body to tackle cyber warfare soon"*, Asian age.com, June 6, 2018.

[412] *Cyber Conflict and Geopolitics—the Cold War's New Front,* Victor Hvozd, Diary of Economic Forum in Krynica Zdrój (Poland) themed "A Europe of Common Values or A Europe of Common Interests?", September 4-6, 2018, Borysfen Intel.

[413] *Beijing Will Give You Cold War Nostalgia,* Walter Russell Mead, November 5, 2019, the Wall Street Journal.

[414] Ibid.

[415] *Cyber War,* Richard Clarke and Robert K. Knake, 2010, HarperCollins.

[416] Ibid.

[417] *Deceiving the Sky: Inside China's Drive for Global Supremacy*, Bill Gertz, 2019, Encounter Books, New York—London.

[418] Ibid.

[419] Ibid.

[420] Ibid.

[421] *The Perfect Weapon,* David E. Sanger, 2018, Crown Publishing Group, Penguin Random House LLC, New York .

[422] Ibid.

[423] *Cyber War,* Richard Clarke and Robert K. Knake, 2010, HarperCollins.

[424] *Confront and Conceal: Obama's Secret Wars and Surprising Use of American Power*, David E. Sanger, 2012, Random House LLC.

[425] *The Perfect Weapon,* David E. Sanger, 2018, Crown Publishing Group, Penguin Random House LLC, New York .

[426] *Deceiving the Sky: Inside China's Drive for Global Supremacy*, Bill Gertz, 2019, Encounter Books, New York—London.

[427] *The Perfect Weapon,* David E. Sanger, 2018, Crown Publishing Group, Penguin Random House LLC, New York.

[428] Ibid.

[429] *Deceiving the Sky: Inside China's Drive for Global Supremacy*, Bill Gertz, 2019, Encounter Books, New York—London.

[430] *Cyber War,* Richard Clarke and Robert K. Knake, 2010, HarperCollins.

[431] *The Perfect Weapon,* David E. Sanger, 2018, Crown Publishing Group, Penguin Random House LLC, New York.

[432] Ibid.

[433] *Was Obama Too Soft on Russia?,* Jeffrey Goldberg, February 15, 2017, The Atlantic.com

[434] *Syria War Stirs New U.S. Debate on Cyberattacks,* David E. Sanger, February 24, 2014, The New York Times

[435] *Deceiving the Sky: Inside China's Drive for Global Supremacy,* Bill Gertz, 2019, Encounter Books, New York—London.

[436] Ibid.

[437] Ibid.

[438] Ibid.

[439] *The Perfect Weapon,* David E. Sanger, 2018, Crown Publishing Group, Penguin Random House LLC, New York.

[440] Ibid.

[441] Ibid.

[442] *The Quantum Computing Threat to National Security,* Arthur Herman, November 11, 2019, The Wall Street Journal.

[443] Ibid.

[444] Ibid.

[445] *Cyber War,* Richard Clarke and Robert K. Knake, 2010, HarperCollins.

[446] Ibid.

[447] *Cybersecurity and Cyberwar: What Everyone Needs to Know,* P.W. Singer and Alan Friedman, 2014, Oxford University Press.

[448] Ibid.

[449] Ibid.

[450] Ibid.

[451] Ibid.

[452] *Cyber War,* Richard Clarke and Robert K. Knake, 2010, HarperCollins.

[453] Ibid.

[454] Ibid.

[455] *Cybersecurity and Cyberwar: What Everyone Needs to Know,* P.W. Singer and Alan Friedman, 2014, Oxford University Press.

[456] *The Fifth Domain,* Richard A. Clarke and Robert K. Knake, 2019, Penguin Press, New York.

[457] Ibid.

[458] *The Weaponization of AI and the Internet: How Global Networks if Infotech Overlords Are Expanding their Control Over Our Lives*, Larry Bell, 2019, Stairway Press.

[459] *With the latest WikiLeaks revelations about the CIA—is privacy really dead?*, Olivia Solon, July 14, 2017, The Guardian.

[460] *The Perfect Weapon,* David E. Sanger, 2018, Crown Publishing Group, Penguin Random House LLC, New York.

[461] Ibid.

[462] Ibid.

[463] Ibid.

[464] Ibid.

[465] Ibid.

[466] Ibid.

[467] Ibid.

[468] *How Americans have viewed government surveillance and privacy since Snowden leaks*, A.W. Geiger, June 4, 2018, Pew Research Organization FACT TANK News in Numbers.

[469] *The Perfect Weapon,* David E. Sanger, 2018, Crown Publishing Group, Penguin Random House LLC, New York.

[470] *The Weaponization of AI and the Internet: How Global Networks if Infotech Overlords Are Expanding their Control Over Our Lives*, Larry Bell, 2019, Stairway Press.

[471] *Complacency Here, Globally Extends Tech's Privacy Reach*, Larry Bell, February 11, 2019, Newsmax.

[472] *The Weaponization of AI and the Internet: How Global Networks of Infotech Overlords Are Expanding their Control Over Our Lives*, Larry Bell, 2019, Stairway Press.

[473] *The truth about smart cities: 'In the end, they will destroy democracy*, Steven Poole, December 17, 2014, The Guardian.

[474] *Who Can You Trust? How Technology Brought U.S. Together and Why It Might Drive U.S. Apart*, Rachel Botsman, (2017) Penguin Books.

[475] *A Global Tech Backlash,* Christopher Mims, October 27-28, 2018, Wall Street Journal, Technology.

[476] *Ex-Google engineer who worked on China search engine calls out 'wrong'*, Daniel Howley, September 19, 2018, Yahoo Finance.

[477] *Fear Mark Zuckerberg's Illiberal Impulses,* Kevin D. Williamson, April 25, 2019, Wall Street Journal.

[478] *A Global Tech Backlash,* Christopher Mims, October 27-28, 2018, Wall Street Journal, Technology.

[479] *Google's Health Deal Spurs Inquiry Into Privacy of Data,* Rob Copeland and Sarah E. Needleman, November 13, 2019, the Wall Street Journal.

[480] *"A Global Tech Backlash",* Christopher Mims, October 27-28, 2018, Wall Street Journal, Technology.

[481] Ibid.

[482] *Sri Lanka's social media shutdown illustrates global discontent with Silicon Valley,* Tony Romm, Elizabeth Dwoskin and Craig Timberg, April 22, 2019, The Washington Post.

[483] *Fear Mark Zuckerberg's Illiberal Impulses,* Kevin D. Williamson, April 25, 2019, Wall Street Journal.

[484] Ibid.

[485] Ibid.

[486] *The Perfect Weapon,* David E. Sanger, 2018, Crown Publishing Group, Penguin Random House LLC, New York.

[487] Ibid.

[488] *Cybersecurity and Cyberwar: What Everyone Needs to Know,* P.W. Singer and Alan Friedman, 2014, Oxford University Press.

[489] Ibid.

[490] Ibid.

[491] Ibid.

[492] Ibid.

[493] Ibid.

[494] Ibid.

[495] Ibid.

[496] Ibid.

[497] Ibid.

[498] Excerpts from Wall Street Journal August 8 review of "The Fifth Domain" by Richard A. Clarke and Robert K. Knake.

[499] *The Perfect Weapon,* David E. Sanger, 2018, Crown Publishing Group, Penguin Random House LLC, New York.

[500] *Cybersecurity and Cyberwar: What Everyone Needs to Know,* P.W. Singer and Alan Friedman, 2014, Oxford University Press.

[501] *Declaration of the Independence of Cyberspace,* John Barlow, February

8, 1996, Electronic Frontier Foundation, Davos, Switzerland.
[502] *Cybersecurity and Cyberwar: What Everyone Needs to Know,* P.W.
Singer and Alan Friedman, 2014, Oxford University Press.
[503] Ibid.
[504] Ibid.
[505] Ibid.
[506] Ibid.
[507] Ibid.
[508] Ibid.
[509] Ibid.
[510] Excerpts from Wall Street Journal August 8 review of *"The Fifth Domain"* by Richard A. Clarke and Robert K. Knake.
[511] Ibid.
[512] *The Fifth Domain,* Richard A. Clarke and Robert K. Knake, 2019, Penguin Press, New York.
[513] Ibid.
[514] Ibid.
[515] Ibid.
[516] Ibid.
[517] *Cybersecurity and Cyberwar: What Everyone Needs to Know,* P.W.
Singer and Alan Friedman, 2014, Oxford University Press.
[518] Ibid.
[519] Ibid.
[520] *Cryptocurrency Exchange In South Korea Reports Theft,* Eun-Young Jeong and Steven Russolillo, November 29, 2019. Wall Street Journal.
[521] *Cybersecurity and Cyberwar: What Everyone Needs to Know,* P.W.
Singer and Allan Friedman, 2014, Oxford University Press.
[522] *Cyber War,* Richard Clarke and Robert K. Knake, 2010, HarperCollins.
[523] *Cybersecurity and Cyberwar: What Everyone Needs to Know,* P.W.
Singer and Allan Friedman, 2014, Oxford University Press.
[524] Ibid.
[525] *The Perfect Weapon,* David E. Sanger, 2018, Crown Publishing Group, Penguin Random House LLC, New York.
[526] Ibid.
[527] *Cybersecurity and Cyberwar: What Everyone Needs to Know,* P.W.
Singer and Alan Friedman, 2014, Oxford University Press.

[528] Ibid.

[529] Ibid.

[530] *DOD's Cyber Strategy: Five Things to Know,* Katie Lange, October 2, 2018, The U.S. Department of Defense.

[531] *The Fifth Domain,* Richard A. Clarke and Robert K. Knake, 2019, Penguin Press, New York.

[532] *The Perfect Weapon,* David E. Sanger, 2018, Crown Publishing Group, Penguin Random House LLC, New York.

[533] Ibid.

CPSIA information can be obtained
at www.ICGtesting.com
Printed in the USA
LVHW112200230320
651002LV00001B/147